Praise for Deviced!: Balancing Life and Technology in a Digital World

"In a world of conflicting reports telling us technology is either wonderful or wicked, healthy or harmful, *Deviced!* is a total game changer. Especially for parents. How do we help our kids—and ourselves—use devices to harness all the advantages technology has to offer our modern lives, while also being savvy and wise about the pitfalls and long-term effects? Doreen Dodgen-Magee banishes our media-elated guilt in favor of teaching intentional and actionable practices that enhance our relationships with each other to carry us through this era and beyond. Highly recommend!" —**Beth Woolsey**, writer and popular parenting blogger of "Five Kids Is a Lot of Kids"

"As technology advances at breakneck speed and we, with our fitness trackers and smart watches, measure and record every step of its progress, Doreen is paying careful attention to the deeper ramifications. With the keen eye of a researcher and the warm heart of a friend, parent, and teacher, Doreen shares invaluable insight into life in the tech age. Her goal in this book is not to call us back to calculators and rotary phones, but rather to remind us of the value of 'fiery, embodied lives'—lives that use technology for our benefit without stunting our own relational and developmental growth. This book is an invaluable tool as all of us wrestle for control over the most important spaces in our lives, the spaces we share with each other." —**Tracy Balzer**, author of *Permission to Ponder: Contemplative Wisdom for the Spiritually Distracted*

"Dodgen-Magee does what most who have written about the impact of technology on youth and adults haven't been able to do—recognize and help us sort out the complexity that technology has introduced into our lives. Rather than build a case for the 'evils' or the 'gifts,' she acknowledges and explains that what we do with technology in our lives is complex and cannot be simply designated as good or bad. Dodgen-Magee provides compelling and useful ways to think about how we might more consciously engage or not engage with technology. This, however, is not a simple self-help book. You will finish the book understanding yourself better because you have explored the intricacies of what you value and what makes you human." —**Candyce Reynolds**, PhD, chair, educational leadership and policy, professor, postsecondary, adult and continuing education, Portland State University

"*Deviced!* is a work of hope and compassion. In this rare book, Dodgen-Magee asks us to notice the altered aliveness devices offer and to consider wholesome alternatives, breaking ground for a middle path many of us seek. Doreen reminds us that we're humans and articulates our innate yearning for play, wonder, love, longing, and a delight in the experience of being alive. This work is unique in not just cataloging tech disruptors but in offering a hand and guidance toward the healing needed to set and monitor attachment boundaries between tech and self." —**Jon Male**, founder, Mindfuls

"Dodgen-Magee, a world authority on the topic of the rational and safe use of the Internet, gracefully reminds us that technology can serve rather than enslave us, that we need to be mindful not to let engagement with devices steal the richness and joy of life from us and our loved ones. Through her creative 'Take Action' sections in the book, she teaches us that an act as simple as eye contact, safe touch, or respectful dialogue can ground us, open channels of communication, and build bridges of trust, so we can continue to remain human even when deviced. In other words, we don't have to turn into machines or check out when we log in! It is indeed a choice to be human. *Deviced!* serves as a much-needed and timely guide as to how to do just that in our increasingly digital world." —**Omar Reda**, MD, psychiatrist and trauma expert, founder of the untangled model of psycho-social care for refugees and trauma survivors

"This is a much-needed book for our time. Dodgen-Magee presents a balanced, nuanced, and compassionate view of the impact devices have on our development. This book is not only about our use of devices; it is about our shared humanity. Doreen calls all of us to engage technology thoughtfully and reflectively so we become more relationally connected and more human. You will come away challenged and enriched." —**Todd W. Hall**, PhD, professor of psychology, Biola University, chief scientist, The Connection Culture Group

"Dr. Dodgen-Magee extensively explores the impact of digital culture on individual lives in a compassionate and humble way. She approaches the subject with a genuine care for her reader while remaining honest to the real dangers of losing one's embodied self to digital spaces. I love this book because it focuses on mending, not condemning, the technology

user to regain an authentic and relational experience in today's society."
—**James Haendiges**, PhD, associate professor, Dixie State University

"*Deviced!* is an essential book for all of us navigating the world of technology. Doreen carefully guides the reader to look at the research, the wonders, the dangers, and the opportunities that exist in the digital world. . . . This book unfolds the story of the digital age and gives us information on how technology has impacted all aspects of our lives including our personal relationships, our ways of entertaining and soothing ourselves, and the way our brains receive information. But this is not a doom-and-gloom book. Doreen celebrates the things that technology is bringing us, and she invites us to embrace technology with moderation." —Lynea Gillen, co-founder, Yoga Calm for Kids, award-winning author of *Yoga Calm for Kids and Good People Everywhere*

"Oh, my goodness! Grab this book now! The introduction alone will inspire you to want to share *Deviced!* with everyone you care about. . . .This is the first book to profoundly transform my understanding of Big Tech's deep impact on a brain-changing, cellular level, to offer practical assessments of what's healthy and what's not when it comes to device engagement, and to propose mindful solutions to tech overuse. As Dodgen-Magee describes, we have come to rely so much on our devices to entertain, soothe, stimulate, and distract us that we miss out on opportunities for deep, authentic, face-to-face connection and for building and strengthening life skills—something critical to consider especially when it comes to our precious children." —**Kathy Masarie**, MD, pediatrician, life coach, and author of *Face to Face: Cultivating Kids' Social Lives in Today's Digital World*

"Researched and thorough, *Deviced!* examines both the benefits and challenges of device-filled living, taking a close, hard look at the costs of technology on both our internal and relational lives. Doreen's secret sauce in *Deviced!* is sharing a brilliant array of action steps for turning inward and feeding our embodied selves." —**Elsbeth Martindale**, PsyD, clinical psychologist and author of *Things to Know Before You Say "Go"*

Deviced!

Balancing Life and Technology in a Digital World

doreen dodgen-magee

ROWMAN & LITTLEFIELD
Lanham • Boulder • New York • London

Published by Rowman & Littlefield
A wholly owned subsidiary of The Rowman & Littlefield Publishing Group, Inc.
4501 Forbes Boulevard, Suite 200, Lanham, Maryland 20706
www.rowman.com

Unit A, Whitacre Mews, 26-34 Stannary Street, London SE11 4AB

British Library Cataloguing in Publication Information Available

Library of Congress Cataloging-in-Publication Data Available

Library of Congress Control Number: 2018950973

ISBN 978-1-5381-1584-8 (cloth: alk. paper)
ISBN 978-1-5381-1585-5 (electronic)

∞™ The paper used in this publication meets the minimum requirements of
American National Standard for Information Sciences—Permanence of Paper
for Printed Library Materials, ANSI/NISO Z39.48-1992.

Printed in the United States of America

For Thomas, Connor, and Kaija.
Thank you for being
my most important teachers
and for indulging this
wild experiment called life
alongside me.

Contents

Acknowledgments

Some of you will notice yourself in these pages. When you do, I hope you will hear the respect and care I have for you. Perhaps you are a college student who stayed after a lecture and pierced my heart with your earnest desire to live a less distracted life, or maybe you've sat on my couch and played with shape blocks while sharing time. It's likely you are a person who has told me a story about how the internet has either saved or harmed you. Maybe we spoke late at night in a dorm hall or early in the morning at a continuing education workshop; or maybe, exhausted and racing, you stopped by during grand rounds and shared your frustration with adapting to new technologies in the workplace. To every precious one of you, thank you! I am honored to hold your stories and hope I have shepherded them well here.

The idea of a fiery life actually began in a home with parents (Fr. Ignatius and Linda) who let my insanely creative and deeply beloved brother, Jeremy, and me move the furniture out of the living room to have a "Hawaiian beach" party, encouraged shanghaiing our friends for surprise breakfasts, and modeled radical hospitality! Jeremy, Mom, and Dad, I have insufficient words to thank you for starting a fire in my life and heart. The fiery life began back then and has continued in the sanctuary home I have been fortunate enough to create with Thomas, Connor, and Kaija. Thomas, I am indebted to you for stretching your funny into my serious and for your undying partnership. Your respect/love feeds me. Connor and Kaija, thank you for being adamant about being your authentic selves (whom I love), and for being both authoritative teachers and gracious students regarding technology and embodied living. You each blow me away, and getting to be your mom is my greatest gift. Ruthie, you already make the fire brighter; welcome to the family! Judy, Ethan, Ella, Gage, Joseph, Andee, Elizabeth, Kindra, Aaron, Annie, Lindsey, Molly, Conner, Ben, Matthew, Katie, Rachel, Chris, Jenna, Taylor, Mackenzie, Mark, Wesley, Katy, Emma, Audra, and Tanner, as well as the Community of In-Betweeners (Tyler, Julie, Ashley, Cassie, Marty, Kaya,

Emma, plus) and all the other "extras" who grace my heart and home with your presence—you play a huge role in my belief that embodied, fiery lives are possible! Thank you for playing with all the toys I leave around, for making me feel smart about doing so, and for inspiring me to new levels of embodied experience. (Amy, I will never be able to thank you enough for gifting me with things I never knew the world or I needed!)

Thank you, too, to Nancy Duvall, for holding belief in me and my intellect in ways that made research and writing possible, and to Vana O'Brien for helping me own that belief in new, challenging, and beautiful ways. Huge thanks also to Jen and Katie, who offered space for writing, and Bruce, Marta, Deanna, Kathy, Judi, Jack, Thomas, Candyce, Tracy, Eric, and Jeff, who provided valuable and specific feedback as the writing commenced. And to Joseph Tatum, "Hey" (which means, "Thank you for perfectly designing everything I need!").

Without a doubt, no one deserves more thanks herein than Ruthie Matinko-Wald, whose expert editing and always sparkling, wise, and spot-on guidance made these pages what they are. If producing a book really is like having a child, then Ruthie is my co-parent and I am the most fortunate one around.

Introduction
How to Read This Book

Full disclosure: when I read books, I usually skip everything before the first chapter. "Stop with the prefaces, forewords, and introductions already!" I think as I flip to the *good* stuff. While you may feel the same and be tempted to skip this, I promise that it is *important* stuff!

I have resisted writing this book for years. As a psychologist devoted to helping people live bold, embodied lives, I have a passion for inspiring people within a shared physical space. For a long time, I pursued this passion in an office with individual clients whose work was inspirational and deep. To manage a growing wait list, I began offering classes and group sessions based on topics such as parenting and developing a healthy sense of self. The unique dynamism of a gathering of people interacting with information—and learning from and supporting each other—was notable. Enthused by these early communal experiences, I began to take every opportunity to engage groups around topics that might help them grow. Putting people and meaningful information in a room together became a favorite activity of mine. Simply by being together, we pushed one another to engage the material more deeply, and profound levels of change in individuals and families took place.

During this same time, my husband was employed in the high tech industry and our children and their friends were getting cell phones. My life among these early technology adopters and influencers allowed me to witness the rapid and full-scale embrace of technology in work, education, and personal spaces. I also was experiencing firsthand the way in which play and interaction among children changed as digital devices entered their worlds. Combining these observations and experiences with my doctoral training in human development, I found myself keenly interested in exploring how technology might impact healthy developmental trajectories for individuals, families, and communities.

As engagement with technology and digital spaces became an ever-growing part of my life and the lives of everyone around me, I began to confirm that, while beautiful and full of potential, tech engagement had the capacity to be a major disrupter to the development of self and social relatedness. So convinced was I that anything short of a moderate embrace of technology would have a significant impact on human development that I started looking for research. In those early days, however, well constructed academic research on the topic was hard to find. I began asking my own questions of just about any "expert" willing to talk. At the same time, I developed patterns of keeping in touch with technology trends and culture's adoption of them.

Coalescing the results of the emerging research, anecdotal information I was gathering, and themes in pop culture, I put together a talk for a local elementary school about how parents' willingness to set intentional norms around technology use might set the stage for healthy development. Like a match to dry tinder, the message took. In the fifteen years since, with no marketing effort, I've been honored to engage with thousands of audiences about how digital devices are shaping us in positive, negative, and neutral ways. I spend, on average, fifteen hours per week culling academic journals, chasing down information on high tech trends, and staying current with popular culture themes. I also spend as much time as possible talking with people of all ages, ethnicities, and cultural groups about how experiences in digital spaces are shaping them. Basically, I start conversations, I listen and research, and I try to provide reliable information to keep the conversations going about being "deviced."

Which brings me back to why I have resisted writing this book. I became a psychologist because I prefer to share information within a shared physical space more than I prefer to write about it. To that end, I have traveled from boardrooms to PTA meetings to university lecture halls to spiritual retreat centers across the United States and abroad. I have led auditoriums of middle schoolers through mindfulness meditations and introduced kinetic sand and pipe cleaners as fidget toys during talks to executive staff members of large corporations. I have gotten on my knees and asked the forgiveness of young adults for my generation's smack talking about them and have begged parents to play video games with their children. I've also intentionally learned how to use devices and mindfully engage their ever-changing technology. I've texted and tweeted ideas for living more embodied lives and blogged about current trends. I've posted well constructed and reviewed research on social media to encourage more mindful tech use. And more. Even still, those with whom I am honored to share the message often ask, "Where is the book? Where can I find more about the impact of being 'deviced?'" So, here it is, with one caveat: by the time this book is in your hands, the technologies referred to within will have changed and morphed and advanced beyond what I can currently imagine or ever hope to describe here. I have decided to publish the book anyway, knowing there are some principles, talking points, and considerations that transcend the specifics of what technology is hot this minute or being released next week.

The bottom line is this: technology and digital devices are here to stay. Oversimplified, all-or-none approaches to understanding or engaging digital devices encourages avoidance of the kinds of thoughtful considerations or meaningful discussions that might make our use healthier and less harmful. Saying that technology is all bad or all good does nothing to address the reality that it is beautifully potent and rife with both positive and pain-inducing potential—with virtue and vice. Denial of its presence or blind constant use can be similarly harmful. When a parent, therapist, or physician says to me, "I don't believe in texting" or "I have no idea what [insert the name of the newest hot app] is," I feel a deep sense of sadness and frustration. This is not completely unlike the sadness I feel when a college student tells me, "Playing violent video games has no effect on me," or a high school student says, "I don't believe in making or answering phone calls."

"Not believing in" things that absolutely exist and are profoundly shaping us is no longer an option. As humans who share space in this world and who are impacted by technological advances every day, it is our responsibility to understand the complexity of the issues at hand. Therapists or physicians not knowing the digital landscape their clients and patients are swimming in is akin to their not knowing about family dynamics or public health risks. Parents not dipping their toes in the digital waters their children are birthed into is, in my opinion, neglectful at best. Similarly, gamers or texters unwilling to consider data regarding the impact of their use are not immune from effects. To be healthy as individuals and communities, we benefit by being willing to look at the issue of how technology is impacting us from all sides and doing so with a measure of grace and openness. This will allow us to use the best of technology in ways that maintain our health as individuals and communities—and to seek ways to curb the kind of use that may harm us.

This does not mean that being open to such complexities is easy. Much difficulty occurs because, in discussing technology, wild assumptions often guide our discourse. It's easier to rattle off opinions than to search for actual data. It's human nature to assume one's assumptions are accurate rather than to tolerate the complexity of new information and different opinions. It's most comfortable to assume that my experience and insights are "right" and those that differ are "wrong." Binary thinking is unconscious and automatic.

Having invested more than a decade initiating and engaging in these conversations with extremely diverse groups of humans, however, I can say that there are few all-or-none assumptions that can be safely applied to our considerations of the impact of our technology use. Therefore, I encourage you to read this book with an open mind and heart. Rather than reading through the lenses of your preexisting biases—which will lead you to either nod in strong approval (passing the book on to your friends and neighbors) or shake your head (tossing the book aside without much thought)—I encourage you to remain open to all the nuance contained within. To be sure, there are positives and negatives around every digital corner; ways that technology can enhance our functioning or disrupt our development abound. The topic of technology is too important for indulgence in binary thinking!

The greatest gift my years of study, travel, listening, and speaking have given me is the repeated opportunity to enter the complicated, messy, beautiful, difficult, amazing, and multidimensional lives of the humans with whom I share space on this Earth. While I enter grand rounds with physicians, ballrooms with psychologists, or university lecture halls with young adults to talk about how technology is shaping our brains, our relationships, and our very sense of self, what always ends up happening is a small or large crisis of consciousness. We quickly move from talk about research and personal tech engagement to discussions about how we are spending our resources (time, energy, money, etc.) and contributing either to a healthier, more empathic and integrated world or to a distracted, contentious, and less healthy one. I'm not kidding: Every. Single. Time. I come ready to present a lecture on technology, and the room turns toward how humanity has become more judgmental and sectarian, or a student wants to talk about how new habits would improve her life, or someone tells a beautiful or painful story about how digital spaces have changed him. I have yet to leave a place without engaging some bigger and more life-encompassing theme than digital medical records, tablets in the classroom, Facebook, Twitter, Pinterest, or *Grand Theft Auto 5*. Yet, since these are the places where we currently spend a bulk of our time, these are the spaces where we will see both our own individual and cultural health—or lack thereof. And these are the places where the conversations can begin.

Consider the well respected mental health physician who brought his son to see me for a consultation, certain the son's gaming was preventing family cohesion. We quickly discovered that the physician's own constant engagement with his phone set the family norm. Turn toward the college students who seek me out between talks to ask about strategies for cutting back on porn use. They want to reverse the physiological effects their use has created and experience greater intimacy in their embodied lives. Become mindful of the parents of young children who also share their struggles with digital demons. Lonely and isolated, they find themselves utterly addicted to social networks and message boards; others find their anxiety is triggered by the constant monitoring of their children via digital means. And then there are the legions of young adults who want to connect about the unresolved fallout from sexual assaults, harassment, or unwanted advances they have suffered during encounters initiated via online dating sites. In this same age demographic are the young couples who feel their relationships are stunted because of a partner's nearly full-time engagement with video games or social networks.

On the other hand, parents have written to tell me how technology has saved their child's life, offered an autistic child the opportunity to know and express his emotions, or allowed an ailing child to be present in school via the amazing invention of screen sharing. Other people have shared how an online course gave them access to training they would otherwise never be able to pursue, thus entirely changing their professional trajectory or their feelings about themselves. Medical and therapeutic services accessed via teleconferencing provide new lifesaving opportunities, and many advances simply make our lives better or free us up to make a greater impact in the world. Technology is changing our lives for the better in an abundance of spaces.

All these encounters, both the seemingly negative and the overwhelmingly positive, start out as discussions about technology but end up reaching deep down into explorations about what it means to be human. The digital world, when explored in its complexity, opens a door for us to experience and learn more about the world of ideas, people, and, ultimately, ourselves.

Pema Chödrön, Buddhist nun and brilliant human, recounts an ancient teaching regarding sacred circles: "Wherever we are," she imparts, "we can draw a circle around our physical beings. This is a sacred circle. Whatever comes into it enters to teach us something."

Read that again, this time more slowly. Whatever (or whoever) enters our personal "bubble" is there to be our teacher. This means ideas that are uncomfortable or foreign enter our consciousness to force us to think through our preconceived notions. This describes both critical thinking and active engagement as tools for maturing the self. Only when we are willing to let complex ideas and reliable data enter our circle, regardless of the goodness of fit between those ideas or data and our preexisting preferences, will we become healthy and resilient people.

When I began this work, I did so with a strong bias and supporting set of assumptions. I believed that embodied interaction was always preferable to interaction via device, that social networks could destroy one's ability to develop relationships built upon a whole and healthy self, and that conveniences afforded by technology should largely be pushed against. After letting all those I have interacted with over the course of my work enter my sacred circle as my teachers, many of those early biases have shifted significantly. I can no longer rely solely on my own assumptions because the stories of people and the research of scientists have taught me that the full picture is much bigger and more complex than my simple assumptions. While living from this kind of "open to consideration" space can be uncomfortable, knowing that we can handle challenges to what we hold as "truth" can be comforting. The way we can know how to handle the challenges comes from examining ourselves, from accepting the limitations of our assumptions, from knowing how to gather and evaluate information, and from being able to tolerate others doing the same—even if we come to different conclusions.

SPECIFIC TOPICS TO CONSIDER

With a topic as large and complex as the one at hand, open and flexible analysis allows for an extremely rich discussion. Some things to keep in mind as we consider the effects of technology include the sticky subject of research and statistics as well as the stereotypes pertaining to generational, gender, and social-economic divides.

The Sticky Subject of Statistics

As noted earlier, research about the impact of technology on humanity was difficult to come by just a few years ago. When I embarked on my mission to help my family and clients with tech overuse, the subject was new. Even today, the subject is only a decade or so old, and the quick advance of technology compared to the long, laborious nature of high-quality, peer-reviewed academic research makes it impossible for the traditional research cycle to keep pace. Granted, plenty of high-caliber studies are now being conducted. But when it comes to the topic of technology's impact, by the time the research is published, the technologies or devices at the core of the studies have been eclipsed by newer ones!

The oxymoronic reality is that, while a historically respected model of research cannot keep up with the speed of advances in technology, technology itself is amassing research data of confounding proportions. This means that some of our most robust information about the digital world can be found in data that has not gone through the rigors of the standard proposal-research study-evaluation-report method of peer-reviewed academic research. This creates a heightened need for informed and wise evaluation skills by the consumer. Not only must we be able to understand if research is ethically sound, but we also must be able to discern the reliability and trustworthiness of statistics gleaned in nontraditional ways.

The truth is that anything claiming to be research must be evaluated before it can be relied upon. A study sponsored by a software company that makes and "tests" a particular game and reports findings about that game must be evaluated for accuracy and objectivity around domains other than monetary ones. What is the sample size? Was there a control group? Has anyone evaluated the study or its findings? In my own experience, this kind of evaluation is rare among those who have not been specifically taught the process of research design.

The Generational ~~Divide Myth~~ Reality (It's Complicated)

Just as statistics related to technology engagement are a sticky subject, so, too, is the "myth" of there being a generational divide when it comes to technology. There is a largely held assumption that younger rather than older humans are more prone to use technology in excessive and unhealthy ways. Similarly, there is a tendency to divide the world into digital natives and nonnatives, idealizing one group and devaluing the other. The work of technology and social media scholar danah boyd has gone a long way to address this trend, reminding readers that being a native does not make one guilty of over- or frivolous use of technology. Neither does placement in the "nonnative" camp accurately identify a person as digitally illiterate.

History provides us with ample examples of times when dividing civilizations into native and nonnative groups did horrible and irreversible harm to one or both groups. Reality is too complex to say that things "were better then" or "are better now," or that "this generation does things this way" and "that one did them better/worse/more efficiently/less efficiently." Things simply were how they were then and are how they are now. Romanticizing the past or idealizing the present serves no one and tells a false story.

All generations are impacted by and involved in the changes occurring as a result of technology's place in the modern world. Granted, in times past, the elder generations might have had a head start on engaging new inventions, which gave them the opportunity to understand and model some of the preferred ways of utilizing those inventions, but we are where we are!

The Gender Divide Myth

Just as falsely dividing perceptions of technology by age is easy, gender, too, presents the tendency for misplaced assumptions. Much of the time, people falsely assume that specific platforms engage *only* men or *only* women or that, in general, individuals who identify as male are more involved with technology than those who identify as female. This is not true.

While entire volumes could be written on the topic of gender and technology, as you read this book, I encourage you to set aside your assumptions about the technology use of "men" or "women" and think instead about the technology use of humans.

The Reality of a Socioeconomic Divide

Both in the United States and around the world, a very real digital divide exists between those with monetary wealth and those lacking it. When resources are limited, investment in technology is diminished. This divide does not exist only between socioeconomic groups in the West but is demonstrated on a global scale between developing countries and those with greater access to wealth and digital infrastructure. To assume that everyone has access to technology and the digital spaces it affords is naive at best and dangerous at worst. With growing amounts of our information and educational resources migrating to digital spaces, lack of sustainable access can have a profound impact on the way in which individuals and communities grow, learn, and function. It's one thing to have the privilege to moderate one's technology engagement. It's another thing entirely to lack access.

Inability to access digital spaces for both communication and information gathering has a profound effect on the way individuals connect with the worlds of both people and ideas. The impact of lack of access on the education of low-income youth is particularly important to consider. The Pew Internet and American Life Project found that students in lower-income schools had greater difficulty accessing and interacting with digital resources, a skill necessary for success in today's academic and vocational environments. Furthermore, 54 percent of all teachers said their students had adequate internet access *at school*, while only 18 percent said their students had adequate access *at home*.

Note that I make no value judgments on which way of living is qualitatively "better." It may well be that those with little to no exposure to technology live healthier or happier lives. We don't know for sure. What we do know, however, is that for people living in cultures of relative privilege and power, technology is here to stay. And those hoping to function in relationship to these individuals will lack important resources without commensurate or, at least, increased exposure. This will shape our communities and our global realities. Thus, whenever we discuss the impact of technology, we must have socioeconomics in mind. If we are to find ways of negating a crippling divide, our keen awareness, global thinking, creativity, and attentiveness will be needed.

BACK TO . . . HOW TO READ THIS BOOK

There are innumerable other preconceived ideas and dichotomies of thought that could be addressed before we dig into the subjects on the following pages. Research not being able to keep pace and the generational, gender, and socioeconomic divides, however, are some of the most prominent foundational issues that undergird most of our current discussions about technology and its impact upon us. These illuminate a human tendency to categorize the world, by a process of massive oversimplification, into binary groups into which all ideas and people are placed. When we engage in this kind of labeling, we risk reducing the complexity of people and their ideas. In addition, such compartmentalization often leads us to idealize the groups into which we fit and devalue those with whom we do not identify. This splitting is the source of much conflict and strife in the world. It is also a pattern that makes it easier to completely sidestep complex critical thinking, problem solving, and conflict resolution.

We must begin with ourselves. Whether our development has been smooth or rocky, whether it has been disrupted or enhanced by the presence of technology or other life experiences, we benefit by rethinking the ways in which we are growing. So take stock and consider the contents herein with an open mind and without judgment. Information can surely help us stretch into healthier spaces.

Part I

How Devices Are Impacting Us

1

There's No Such Thing as IRL

In the recent past, the acronym "IRL" has come to stand for the phrase "in real life." It refers "to a person's non-digitally based, fully analog, three-dimensional, in-one's-own-and-actual-skin self." Although the acronym isn't that old, I think it's time for retirement. The reality is that our digital and embodied lives come together to create one real, whole life—a real life that includes both. Actual friends exist in both embodied and digital spaces; they include our clans, classmates, and support groups—even though we may never meet in what used to be called our "real lives." We buy physical items in virtual shops, we learn important lessons and gain actual skills in digital spaces, and we carry in our pockets virtual assistants that often know us better than our embodied friends; when these assistants fail us, we feel real frustration. All of life, including that lived in the digital domain, is real.

I experienced the frustration of the overlapping of digital life and embodied life recently when trying to engage voice recognition to make a call to my friend Maiya. From my first attempt, Siri had huge issues with Maiya's name and how I pronounce it. Upon realizing this, I added "friend" after "Maiya" in my contact list, because every time I asked Siri to call "Maiya," she just froze.

One day I had invested more time than usual trying to get Siri to call Maiya for me. Using voice commands, I enabled my phone and car audio system to sync and then asked Siri to "Call Maiya Friend." "I'm sorry. I can't find that name in your contacts" was the response. Patiently, I tried again. "Here's what I found on the internet for 'My Ah Friend,'" piped a resourceful-sounding Siri. Now frustrated, I countered, "No! Call Maiya Friend!" Again, Siri fed me an internet search for "My Ah Friend." "Fine! I'll call her myself! CANCEL!" I snapped into my car's audio system. Shockingly, Siri actually answered: "I'm sorry I didn't understand that pronunciation. Can you teach me?" I nearly pulled over, I was so surprised. I decided to try. "Maiya Friend," I said with the proper pronunciation. "How can I help you?" responded Siri. "'Maiya Friend' is pronounced 'Maiya Friend,' not 'My Ah Friend,'" I said. Before I knew it, a long internet search appeared on my screen and Siri proudly reported what she had found in her second internet search for "My Young Friend." I sense that we all have similar stories that result in *real* feelings sparked by digital experiences.

TECHNOLOGY AS THE AMBIENT BACKGROUND NOISE OF EVERYDAY LIFE

To act as though this interwoven reality does not exist and that our digital lives are somehow removed from our embodied ones seems silly at best and foolish at worst. For the majority of culture, regardless of personal choice, technology has come to create an ambient background noise that is virtually inescapable (or escapable only by great effort). For those who choose to engage digital spaces volitionally by intentionally creating even small online presences (think social network profile, online account of any sort, entering preferences into an entertainment site, etc.), by owning a computer or tablet, or by carrying a digital device, the dual nature of life as lived online and in embodied ways is acknowledged. By simply engaging the medium, users are investing a part of their physical lives in digital spaces and making it such that, regardless of within which domain an action happens, all that person's experiences are part of his or her "real life." In this economy, all of life's occurrences, whether digital or analog, are part of our real lives.

Like it or not, the constant availability and presence of an always-on digital environment is an instigator of feeling and experience, even for those who rarely take advantage of its full potential. My mom, for example, carries a flip phone with no camera feature but loves having photos of the meaningful experiences in her life. She has a small digital point-and-shoot camera but often doesn't have it on hand when she'd like, leaving her dependent on others to email her their digital photos of shared experiences. When she receives these pictures, she often has difficulty scaling them to size to print them. So, yes, while Mom engages technology, she is not tech savvy, and her attempts to use technology for something that matters to her often leads her to feel incompetent and out of touch. The experience also robs her of far more time and energy than she'd like to give. These real feelings and real lost time exist in her embodied life. They leave her aware of a sense of lack and frustration.

On the other end of the spectrum is Jorden. Jorden lives a good chunk of his life in digital spaces, finds most technological opportunities life expanding, and highly regards the digital world. He works as a designer for an app developer, wears an Apple Watch, and engages social media in efficient and creative ways. Desiring to date in a new city, he created a Tinder profile, which he engaged with high levels of discernment. Quickly, however, Jorden found it difficult to turn away from the app. Even when he did, he found himself curious about who might have swiped right on his profile. Over time, he began to notice an ambient type of hypersexualized thought, knowing that he could at any time access the app and find a match. After a few dates, Jorden chose to delete the Tinder app, turning to other methods of finding connections. This eventually led Jorden to reassess his time on Facebook, realizing that the time he was sinking into that online social network was disproportionate to the time he was spending developing his embodied life. Investing long periods of time gazing at photos of friends from his past made him lonely and caused him to idealize what had been. Before completely deleting his Facebook page, he first asked a friend to change and hold the new password so that he could experiment with a week off. Eventually, he pushed "Delete." This was not an easy decision. Choosing to delete his Facebook account meant losing a decade of meaningful investment in a community and a presence that mattered to him. But doing so has helped Jorden feel more actively present in his daily, embodied life, and his current relationships are thriving.

Somewhere in the middle of the spectrum lives Debbie. In her mid-fifties, Debbie went through a major life change with a new single status, home, and degree. Directing her job search mainly via digital means, socializing through online dating sites, learning to take full advantage of her smartphone's many tools and functions, and finding new volunteer opportunities via online networks have all been lifesavers for her. Granted, the learning curve is steep, and Debbie experiences frustrations as she learns new technologies, but the payoffs for her are significant, so she feels the challenges have been worth the effort.

Each of these stories demonstrates the way in which technology has become a part of daily life in varying proportions. Whether the impact is minimal and contained or significant and woven throughout the bulk of our experience, having a constantly available resource surely colors the way in which we are present (or not) in our lives. In fact, our devices are increasingly becoming friends and assistants of sorts that both hold and direct our experiences. We take an inordinate number of photos of our food, pages of books, and random things we love, simply because we can. We read numerous reviews before we purchase an item or frequent a business because such curated opinion is available to us. We track our steps and our sleep, and we talk to our pets and children from afar via remote monitors. None of these actions are inherently bad or negative. Yet they tell a subtle story about how our devices and the spaces they make available to us have become part of our very selves. Over time, we rely on digital devices more—and on ourselves less. We take fewer risks because we rely on the reviews of others. We ask less of our memories because our devices remember for us. And often we don't engage the people and places surrounding us because we have come to prefer the company of devices that ask so little of us in return.

I frequently think of the psychological concept of "leaking" when I consider these realities. Leaking, as I use it here, involves syphoning off just enough internal psychic pressure to make us comfortable staying exactly where we are instead of moving toward newness and growth. If we feel a need, desire, or place of discomfort within ourselves, we realize that meeting the need or removing the discomfort will require work of some kind. When this work is difficult, of unknown nature, or hard to get a handle on, we delay the work—and the unmet need or discomfort builds. As it builds, so does our desire for relief. Rather than engage the process of resolution directly, we look for (or stumble upon) ways to release small bits of the discomfort within us: We shop or eat or drink or smoke. We exercise excessively or tune out online. We find people who will take small bits of our burdens, or we isolate ourselves. We do all these things to let off just enough internal pressure to continue to live with the need or discomfort without fully addressing it. We treat our insides like a boiling pot on the stove, lifting the lid by all manner of behaviors, trying to let off just enough steam that the contents won't boil over—but we never turn down the heat! Let me use a story to explain how this dynamic plays out in relation to our use of technology.

Lauren has always enjoyed a rich inter- and intrapersonal life. He loves learning and has a satisfying social life. He has a variety of friends and interests. Hailing from the West Coast, Lauren took a job on the East Coast with a high tech start-up after he graduated from college. While Lauren finds his job stimulating and his coworkers fun, he misses the slower pace and easy connections he fostered prior to his move. In California he had intellectual, social, and faith communities that both motivated and encouraged him. He was familiar with the owners of the businesses he frequented and felt a deep sense of belonging. In addition, Lauren was an avid surfer, and he now finds his body longing for the familiar rhythm of a week of work followed by two days in the water.

As he moves more deeply into the reality of his present life, Lauren feels a growing loneliness and lack of meaning. Efforts to deepen the connections he has in his new city leave him feeling homesick and motivate him to text and Snapchat his friends far away; he also spends inordinate amounts of time scanning their walls on social media sites. When Lauren grows tired of this, he turns to tide charts and surf reports, dreaming of how he'll manage his vacation time. When all else fails to quell the discomfort within, he turns to his laptop and immerses himself there in either work or endless streams of YouTube videos about surfing.

While none of Lauren's behaviors are inherently bad for him, they do not meet his deeper need: to address his loneliness and sense of being adrift in a new community. Using his preexisting connections via quick messages is a great option for getting through tough moments. When he engages others exclusively in this way, however, it isn't so much for authentic catching up and deepening the relationship. It is more for finding a handhold to get through a painful moment. When done repeatedly, this lets off just enough pressure to relieve Lauren in the moment. The trouble is that this keeps him from feeling and dealing with the full weight of the need to either get himself settled into a new life where he is or return to his prior one. By habitually reaching out to his connections far away, Lauren might not become uncomfortable enough to do the very difficult work of engaging where he is or of making the difficult decision to return to his previously rewarding life. The same could be said of his use of surf reports and work. By distracting himself, he turns away from the internal pressure of his feelings until he no longer feels motivated to do the work of making significant changes.

Most of us can relate to this kind of experience. We don't deepen our relationships with our devices or digital spaces with real intention. Instead we do so almost accidentally. Over time, these deepened connections end up serving us well when we feel a need or sense of discontentedness in our embodied lives. We thus feed a vicious cycle of falling more deeply in step with our devices and losing opportunities to stretch and develop our embodied selves. This results in less actual embodied experience with ourselves and our surroundings, which leads us to feel as though we belong more to our devices and less to our embodied lives. All these small "leaks" of pressure off the boiling pot rescue us from developing what we need in our physical bodies and intrapsychic selves in order to deal with the "heat" of embodied life.

CONVENIENCE AND COMFORT

A participant at one of my talks many years ago spoke of seeing her job as a therapist and parent as being one of helping people learn to tolerate being inconvenienced and uncomfortable. This sentiment has made a profound impact on my thinking, and I have come to believe that it speaks to one of the greatest tasks before us. In a world of constant access to devices that offer us convenience and comfort, might it be our work to determine both the optimal amount of each and what amount will end up harming us?

The tech industry's goal in most things digital is to offer the consumer both convenience and comfort—not bad offerings in themselves. In optimal doses they help us and allow us to be productive, content, and available to life. If, however, we exist in exclusively convenient and comfortable spaces, we lose appropriate motivation to undertake the kinds of risks and experiences that keep us growing and maturing. The dynamic is similar to that of anxiety; a healthy and moderate level of anxiety can help us perform. As a public speaker, I know well the truth that an optimal level of anxiety is necessary to perform at my peak. Too much and I am paralyzed—likely to be stiff, struggle with appropriate recall, and lack a flexibility and warmth that adds credibility to my presentation. Too little and I simply don't show up. I underprepare, under present, and underwhelm.

The same can be said of convenience and comfort. Too much of these two commodities and we lose our edge. We stop feeling the nudge to persist and engage meaningfully. We come to feel entitled and bored in all the worst ways, pursuing hedonistic and narcissistic pleasure and validation. Too little and we are not free to experience the joy and freedom of spirit with which we were meant to live. When conveniences are engaged to allow ourselves time to pursue experiences that will help us grow, benefit person-kind, and expand our horizons—fine. When they are engaged solely to help us live with too much ease, avoid responsibilities, stay in comfort zones, and free up increasing amounts of time for leisure only—not so much. The trick is to find the fine-line balance, to feel convenienced and comfortable enough without becoming too much so.

2

This Applies to You
(and Me and Every Other Person)

Technology is ushering in a spectacular range of cultural advances. Previously isolated people have access to life-enhancing resources and global communication, information reach has never been better, and medical and scientific capabilities advanced by technology are saving lives and the planet. Rich digital environments provide enhanced learning opportunities, and assistive technologies are saving the day for many individuals. Technology truly is a grand reality, and the digital world has many gifts to offer.

But alongside the gratitude for the ways that technology benefits society, there is a real and growing sense that in terms of day-to-day use, few people *under*utilize or even moderately engage it. Having rushed to adopt both the truly amazing advances and the banal conveniences, many are experiencing niggling questions about how a near-constant engagement with devices might be having an impact on certain domains of human functioning. Complicating things, while we are waking up to an awareness that our excessive technology use may not be benign, we are in many ways increasing our full-scale dependency.

Wherever I go, in fact, I encounter people who are concerned about humanity's overreliance on technology. The retirement set is frustrated that their grandchildren no longer answer ringing phones, and young adults wish their grandparents would text. Parents complain about their children's gaming and social networking. Adolescents report frustration about the hypocrisy they experience in their parents' complaints, noting the challenge of getting the adults in their lives to look away from their own devices. Single and partnered people, the young and the old, and everyone in between speculate about the possible problems that could result from a seemingly universal full-scale adoption and passive endorsement of the digital experience. In addition, academic, educational, and medical research studies are uncovering changes to brains and behavior patterns when technology use is high. Teachers and professors bemoan shortened attention spans and increased distractibility in the classroom and point to digital devices as the cause.

Obviously, technology's advances usher in both positive and negative consequences. This was starkly illustrated for me recently while doing grand rounds with medical students. I heard repeatedly about the conflict presented by the computer in the exam room. The assembled students' motivation to become physicians was initiated by a shared desire to encounter people and help them move toward health. Their training included much instruction to this end. In their applied residencies, however, much of their experience was figuring out how to truly be with a patient while at the same time fulfilling the medical system's need for a thorough and hyper-timely digital record of each appointment. According to the residents, computers—installed in such a way as to hover almost directly between the physician and patient—felt like a constant, tangible presence in the room that demanded the kind of attention that had previously been dedicated solely to the patient. While providing access to increased helpful data, they also impacted the fullness of the personal encounter.

Later, at an appointment with my own physician, we talked about this anecdotal finding. He shared that the doctors in his office who have the highest digital charting compliance scores have the lowest patient-satisfaction scores. Those who have the highest patient satisfaction scores have some of the lowest digital-recordkeeping scores in the clinic. While one might argue that this dichotomy will diminish as physicians and patients become accustomed to the presence of digital devices in the exam room, important thematic and symbolic realities must be addressed. When a digital device is engaged during an embodied human encounter, what is each party's relationship to the device, and how is the intimacy of the encounter affected? If it is affected, what are we doing to address or control for this? The reality into which we are evolving in many settings is one wherein determining how to handle the potentially disruptive powers of technology is a complex and ever-moving task, often largely outside our personal control.

On college campuses, I am frequently interviewed by university communications staff. Most often these are upper division or graduate students ranging in age from twenty to twenty-six. Prior to our interview, they have either heard a lecture or two of mine or watched one of my talks online. The most common question I receive is one that also stops me short no matter how many times it is asked. The question goes something like this: "You talk about moderation. As you do, I realize this is exactly what most people my age crave. When you ask us to power off our phones and lead us through the ten-minute mindfulness exercise, I sense that we all breathe a collective sigh of relief. You speak of setting strong technology use norms and maintaining them. The problem is: We are hired and taught by people who require us to respond to texts and emails when those texts and emails arrive, regardless of when they arrive. We are being educated in a system in which a significant amount of our time must be spent reading, writing, doing research, and collaborating online. How do you suggest we maintain a healthy balance when so many of the demands for our educational and professional success revolve around technology use and are outside of our control?" I have immense empathy for all people trying to negotiate these complex realities in the workplace and in academics. It is difficult to find ways of managing the preponderance of digital forms of entertainment and communication alongside the realities of 24/7 required work availability, let alone embodied living.

Life used to force opportunities to rest and be bored simply by virtue of the fact that we did not have the capabilities we have today. We are no longer required or naturally provided with opportunities to develop the skills that boredom, inconvenience, and discomfort provide us. We don't always get the luxury of setting healthy norms for ourself while at the same time succeeding in our chosen vocational or academic goals. We may be required to complete a digital medical record while meeting with a patient. We may need to keep our ringer on throughout the night or face dismissal from our job. These are our present realities.

Even still, I call us all to consider and work toward a world in which we can find moderate norms that keep our entire lives and selves healthy and tended to—and that we create realities where our employees, students, and neighbors can do the same. Basically, it's important to try to find ways of moderating our use when we can and ways of honoring others' attempts to do the same. It's important to talk with professors, managers, employers, and friends about how to navigate technology *and* embodiment. It's important to work to determine what health would look and feel like for each of us and to move, in successive approximations, toward some semblance of balance. If none of us push back against our cultural full-scale, technology-for-everything-and-in-every-situation reality, there will be no counterpoint to the full-speed-ahead drive that is inherent in the digital world. In ways large and small, I hope we can all be part of the movement toward moderation even though, and perhaps especially because, it is such a seemingly impossible reality to accomplish.

To my mind, the matter at hand is not technology alone. It is, instead, the way in which we engage technology and set norms around its use. When I first started speaking about how technology was making an impact on human functioning, I tended to idealize the past. I romanticized times when there was a single phone and television in most homes, and when busy signals on phones and the complete shutdown of network television happened every night between 1:00 and 6:00 a.m. Given the conversations I was having with kids who were navigating the early days of cell phones with all their newfound potentials for "good" or "harm," I wished for an easier time.

Once, to indulge this wish a bit, I sat down with my big box of *Life* magazines from the 1950s. There, right alongside the idealized wall-mounted kitchen phones and console televisions I dreamed of, I found a preponderance of advertisements for cigarettes and convenience foods. One ad pictured a physician in his white lab coat recommending a specific brand of cigarette as being helpful for "throat scratch." On the next page was a full-color ad picturing a plate of fried tomatoes, chicken, and doughnuts, claiming that anything fried in vegetable oil was more "healthful for your family."

In the postwar 1950s, Americans were coming out of a time of rationing. Commodities that had previously been lacking were now abundantly available, and marketers were ambitious in their selling of the idea that the time for convenience had come. Mass-produced and marketed, easy-to-prepare foods were a beacon of ease and comfort. Loaded with preservatives and salt to enhance flavors, artificially colored to add appeal, and packaged to be enticing, convenience foods caught on. In short order they went from being the side dishes they were intended to be to becoming mainstays of the American diet. Similarly, cigarettes, seen as a benign indulgence, became highly embraced by the American public. Approximately twenty years after this boom in the making and marketing of convenience foods, the first increases in high cholesterol, hypertension, obesity, and high blood pressure can be found in population health literature. Smoking-related cancer findings also emerge around this same time. Looking back, it seems we were conveniencing and comforting ourselves into highly unhealthy states.

As a group, Americans are not necessarily known for moderation. In the 1950s we were seemingly unconcerned about how a dedication to canned spaghetti or boxed macaroni and cheese might impact our health. We trusted the intentions of those producing these foods and the agencies that governed their production, and we assumed that our best interest was their goal. We found that the time we saved by using these quicker-to-the-table options meant opportunity for other experiences. As we acclimated to small amounts of these flavor-enhanced products, we found ourselves wanting more. A moderate embrace might have saved us the health impacts, but a maximal embrace was what we achieved.

My sense is that a similar dynamic is at play with technology. We engage it a little bit and find it to be "tasty" in ways our embodied lives aren't. We use it to save time here and there, and suddenly we're investing the moments we've saved back into our technology engagement. We get comfortable with our use, only to think that more use might offer us increasing amounts of free time, social connection, learning, or entertainment. Before we know it, we are spending large quantities of our lives in digital spaces. What was meant to be a side dish or accompaniment to our embodied life has, instead, taken center stage. When this happens, we might seem to gain a lot. Sure, engagement with technology has *the capacity* to result in increases in creativity, collaboration, socialization, visual reaction time, and visual spatial awareness. The trouble is that there are plenty of potentials for *less than ideal* consequences as well! Disruptions to our physical bodies as well as to our inter- and intrapersonal lives are well documented.

Throughout our lifespan, we move across a developmental continuum. Along the way we are faced with experiences and opportunities that catapult us forward, push us backward, or interrupt our journey. Both "positive" (e.g., mastering the ability to function as an autonomous self, committing to relationships with people or communities, finding one's passion) and "negative" (e.g., significant loss or rejection, failures, discovery of limitations) life events hold opportunities for forward and reverse movement. Typically, we move in fluid ways, making steps forward and backward and rolling with challenges and obstacles.

When an event causes a significant shift in our developmental trajectory, however, it can be loosely defined as a "crisis." In this use of the word, "crisis" simply refers to the disruption of a preexisting pattern of thought, emotion, or behavior. A career change can be considered a crisis in that it shifts our daily patterns of behavior and thought. The initiation or termination of a relationship causes a similar kind of crisis. Even exciting events such as landing a promotion, completing a task on which we have worked hard, or adding a new member to a family serve as crises in that established patterns and routines as well as thought and behavior patterns shift to accommodate for the changes made externally.

With some of these crises, existing sleep, socialization, and daily life patterns shift to accommodate the newness; in others, changes in physical location or personal expectations shift drastically. All these crises could be called "disrupters" in that they confront us or interrupt us while we are engaged in our everyday journey of development. Often, what keeps crises or disrupters from holding us hostage or moving us backward is our relationship to the particular crisis or disrupter and our commitment to forward movement (recognizing that sometimes significant obstacles are larger than even our desire and effort to move forward). Our character plus the specific nature of the potential disrupter come together to impact the deeply internal process of responding in intentional ways that allow for continual forward movement, mindless response to the disrupter, or movement into a fight/flight/freeze response in the face of the impending crisis.

> **Potential Disrupter + Character/Personality/Developmental Milestones Reached to This Point = Either:**
>
> - Confrontation of the disrupter/working through, allowing for forward movement
> - Mindlessness in light of the disrupter, causing a sort of spinning of the wheels
> - Fight/flight/freeze, causing a developmental arrest

Over the course of my life, training, and professional work, I have come to see technology and digital devices as being a particularly enticing and ever-present disrupter of both daily functioning and ongoing development. The field of interpersonal neurobiology (IPN) provides strong evidence that embodied relationships between caring humans set the stage for robust, complex, and integrated development. The constant presence of digital devices introduces a third party into human relationships in such a way as to alter our relationships with our most basic selves and the "selves" of others.

Infants now wear digital monitors that inform parents of all manner of statistics. Digitally enabled toys store data in the cloud, informing complex algorithms related to the users and shaping further interaction with both devices and information. Think about it: As children we may be playing, learning, and engaging with friends in digital spaces. As adults we may use wearable technologies to inform us of our own bodily functions and rely on devices to connect us to others. While none of these trends are inherently "harmful," they tell the story of lives often connected most devotedly to technologies and embodied connections moderated by digital devices. This reality has complex ramifications for the developmental process, offering opportunities for the disruption of healthy development in several ways across many domains. While these disruptions may present as crises that spur progression, very often they serve as roadblocks of sorts, holding us aside from a healthy growth trajectory. This is especially true when digital devices have been integrated into our lives without forethought, intentional planning, and moderation. Because of this, an exploration of the potential for disruption can have profound effects on how we choose to engage digital domains.

THE REALITY OF THE IMPACT

Common Sense Media, in its groundbreaking studies of entertainment media and time spent in media domains, found that US teens, ages thirteen to eighteen, average approximately nine hours per day engaged with media; tweens, ages eight to twelve, average approximately six hours per day; and our littlest ones, ages zero to eight, average approximately two and a quarter hours per day with screen media. None of these numbers include time spent online at school or doing homework. And it isn't only our children who are spending massive amounts of time with devices. Research by the Nielsen Company found that the typical American adult spends, on average, ten hours and thirty-nine minutes per day interacting recreationally with screens. Given that most Americans now use more than one digital device at a time, the specific number of hours that researchers claim we are spending with screens is actually greater than this; that "real" number would be most realistically calculated by putting the time spent on each device end to end. This means that while one hour may go by, if an individual is on both a cell phone and a laptop for that entire time, the screen time would be recorded as two hours. This goes a long way in explaining why the number of hours we spend with devices has been growing steadily for the past fifteen years, with the accumulated time appearing to be taken from family time, social practice, physical activity, and sleep. The impact of sleep loss and sedentary lifestyles is well documented in accessible sources and will be discussed in depth in later chapters. The effects of diminished family time and social practice are less frequently discussed but equally important.

Family talk time and social practice are two of the ways we learn to be in the world. When we spend time being together and conversing with our primary family and support community, we learn how our presence makes an impact on others and how our words and very sense of being do or do not "land" with them. We learn that certain behaviors lead to peaceful and harmonious interactions and that other behaviors do the opposite. Ideally, in these settings we learn how to communicate, manage conflict, and function in a group. And hopefully we experience that we are loved for who we are.

With social practice we get to find how our personhood impacts a group of others outside our homes. We find out what forms of communication do and do not work, and we encounter opportunities for the expansion of empathy and relatedness. If our family relationships have allowed us to develop a sense of security about ourselves and our abilities and our social practice settings are relatively safe, we can use social practice opportunities to learn how to make mistakes and live through them, how to honor the experiences of those around us, and, generally, how to play well with others.

When opportunities for the two important developmental processes of family talk time and social practice are diminished, we are greatly impacted. Not only are we deprived of the simple pleasures they offer, but we also miss out on commensurate skill-building opportunities. Furthermore, we may feel we aren't really important and we don't really belong, which will affect everything we do and experience.

Unfortunately, tragically at times, most of us aren't aware of what we're missing by having allowed devices to disrupt our engagement in these spaces. In fact, academic, peer-reviewed research as well as less-rigorous studies find that users tend to underestimate the amount of time they spend with technology—never mind the amount of energy and attention we devote to it. Being honest with ourselves about the way we spend our time is uncomfortable, especially if we spend it in ways that allow for the formation of habits that may not be wholly healthy. Few individuals rush to disclose how many grams of sugar, caffeine, or fat they consume in a day. Similarly, our technology use habits are frequently such that we prefer to be unaware. We cringe about how many texts we send and prefer to ignore our serious online time-wasting habits. We fail to seriously consider how many minutes we've sunk into watching multiple episodes of a favorite show or how many hours we've invested in our video game play. We are not consciously aware of how our dedication to our devices might have limited our talk time and social opportunities in our embodied spaces.

Even if we are individuals who claim to have a good handle on our technology use, there is typically some platform that captures time and attention in a way that we underestimate. We are the ones who claim to never be on Facebook but who are always the first to comment on a post, typically seconds after it has been published. We are the ones who pull out our cell phones during dinner to show the hilarious YouTube video we found, focusing the attention onto a screen and away from those bodies at the table. We are the person who loses a day to anxiety about an un-responded-to text. Yes, most of us have our digital vice!

If we don't relate to the above, then we are likely someone who is overly critical of culture for its full-scale technology adoption. If so, then we, too, need the information in these pages to become informed and educated. We need to grow our understanding of the disrupters of our own development, and to increase our empathy toward others whose development may be disrupted by technology. We need this information to help us start non-shaming discussions focused on listening to those about whom we care most.

BACK TO WHY THIS APPLIES TO EVERYONE

I recently purchased a new car. It would be relevant to note that I hate shopping in general and hold a special place on my "highly dreaded activities" list for the procurement of items that cost more than two hundred dollars. These types of purchases require time, critical analysis, and a great deal of information gathering. When shopping for something as significant as a car, a deep sense of groundedness helps. Being able to think clearly, to evaluate and act on relevant real-life data as opposed to relying on personal preferences and biases alone, and to pay attention to both ideological and physical experiences are all important. Buying a car simply because it's the easiest one to procure in my favorite color would be unwise. I need to think and research more deeply in order to purchase a reliable automobile that meets my unique needs, and I resent having to put this effort toward this task!

To do as little in-person shopping as possible, I did a fair amount of research online to narrow my initial search. Within hours of said internet perusal, the two cars I was considering were advertised on nearly every web page I visited. As time went on and I continued shopping, the cars in the ads became more specifically tailored to what I was settling on, reflecting the color and model I was most strongly considering. Before I brought my conscious awareness to the table, I found myself thinking, "Wow! That one is consistently on the front page of the *New York Times*, Slate.com, and ApartmentTherapy.com. That's saying something! I've got great taste, and this car must be absolutely rock-solid reliable to be featured in so many places!"

The way in which my search history was shaping both my online experience and the content of my thoughts surprised me. As a person who studies the impact of our technology use on our embodied lives and who intentionally creates time for quiet reflection and analysis of decisions I make, I would hope that I wouldn't be as easily disrupted. If *I* were allowing my embodied purchase choices to be determined by online ads that I myself, in the form of the algorithm that is "doreen," had created, how much more so are those who haven't thought consciously about these things being impacted?

This gets to the heart of why the things I write about here apply to me and they apply to you. All of us who breathe and move about in this world do so with either an active or passive awareness that we are developing all the time. The constant presence of disrupters catches all of us off guard at times. Only the very dedicated outliers escape this reality. Go, outliers! You will hold the balance for us as the rest of us fight overuse.

Today, technology serves as an enticing distraction on our developmental journey—whether we carry a digital device in our pocket or purse or bemoan those who do. Whether we game obsessively in our free time or only go online periodically, we all coexist in this time and place where much of humanity is being impacted by the growing presence of digital devices and the online domains they deliver.

If the goal of living is to continually grow and mature, we must take a long look at our own development and how it is helped or hindered. The very core of ourselves is affected by the experiences in which we allow or force ourselves to engage. Our minds, guts, and bodies are shaped by the narrow or broad realities to which we expose them. More than ever, we must do this work with intention and by sheer act of will, or our trajectory will be narrow and limited. To avoid complacency, work through developmental arrests, and become healthy and whole, we must examine the nature of our journey, the ways in which we invest ourselves and our time, and the disrupters that influence both.

3

Living Outside Our Skin

Humans are sensual animals. We touch, taste, smell, and see our way through our days. When we are aware of our sensual selves, we pay attention to both the *message indicators* that direct us to stimuli and to the *physical experience* of these stimuli. Our stomachs rumble, so we look for food. We yawn, realize we're tired, lie down to go to sleep, or go outside for a renewing breath of fresh air. We smell something out of the ordinary and search for its source. When we are especially mindful and aware, we may realize that we yearn for a vision of beauty (or something otherwise satisfying) or to be touched in a meaningful way. Each of our senses serves a unique function in keeping us healthy and content. When one or more senses become compromised, we often find the others become heightened to compensate and keep us aware and in tune with ourselves and our surroundings.

As our technology use increases, it isn't surprising that we may fall out of touch with the potency of some of our senses. We begin our day by rolling over to grab our phones to catch up on the news or on our various social "feeds." We connect with our device before we get out of bed or truly wake up to our embodied self! During the day, we surf the web to tell us what to buy; use digital monitors to tell us our child's or pet's body temperature, heart rate, and more; use wearable technologies to track our exercise and heart rate; and stave off boredom by tackling another level of a favorite game while waiting in line at the coffee shop, bus stop, or grocery store. At the end of the day, we plop down on the couch and play a video game, or we crawl into bed and watch a movie on our tablet or laptop—because we're tired. Once in bed for the night, we track our sleep with our fitness bands. As we engage these platforms, we send our minds and bodies the not-so-subtle message that our technologies can entertain, comfort, and know us better than we can entertain, comfort, and know ourselves. This tendency to rely on devices, apps, and technology more consistently than on our own sense of our mental, emotional, and physiological states has far-reaching consequences.

Our very sense of place in the world is similarly impacted. At the core of our consciousness, we no longer truly need to know where we are, as long as we have a device and an internet connection. Directed by our GPS and Google Maps, we go directly from where we are to where we want to go without having to think or wonder. We visit a new part of town or a new state altogether but have no sense of our larger environs. We don't notice the landscape because we're busy following our turn-by-turn directions. We don't interact with the "natives" of these places, and we find our favorite "local" chain restaurants and stores in every place we visit, so we can frequent the familiar rather than brave the unknown. We rarely stop to attend to what it feels like to be an embodied person in a new space, and we don't travel consciously into new spaces. Instead we move through them, looking down at our phones the entire time. When we do look up, we snap photos on our phones, relieving our minds from the need to hold the memories.

DIGITAL TRACKING VS. SELF-KNOWING AWARENESS

While the term "self-knowing awareness" refers to many abilities, it provides a wonderful phrase for considering how well we know our own physical bodies. If we have a developed ability to scan our physical bodies—paying attention to what our sensory awareness can tell us about what we need and prefer in the way of stimulation, rest, learning, and more—we can trust our own internal "gut" to inform us how to live most healthfully. If, however, we have lived in such a way that a bulk of our stimulation, soothing, learning, and information gathering has come from *outside* our bodies, we feel bereft of knowledge regarding how to live in healthful ways in and of ourselves. As with so many domains of our functioning, this has much to do with the kind of living we have actively and passively *practiced*.

Human bodies are amazing machines. They are equipped with message indicators that can inform us of just about everything we need to know to survive and thrive. Included in that information can be everything from our energy level to our need for rest, nutrition, connection, and more. Think about it: yawn = being tired; stomach growl = hunger; tears = sadness or joy; fast heart rate = overexertion, anger, excitement, or medical anomaly. To inventory and assess what our body might want or need, however, requires a *practiced* pattern of checking in with ourselves, often in quiet and still ways. When we outsource this process to external devices such as monitors, apps, software, and the like, we miss out on the opportunity to know ourselves deeply and to practice self-regulation.

Our cultural obsession with fitness trackers provides some insight into how we tend to favor attendance to digitally tracked data over our felt, lived experience. These devices, which record movement throughout the day, can also be worn to track and evaluate sleep. Interestingly, most of these devices monitor sleep through motion detection, based on the premise that stillness correlates with sleep and movement with wakefulness. Analysis of clinical sleep studies done at the same time as sleep monitoring with fitness trackers, however, reveals a great degree of variance in the accuracy of sleep assessment via wearable devices. Sleep, with its various stages and cycles, is a complex activity. While movement is one indicator of depth of sleep stage, many other variables contribute to its nature and quality. Nonetheless, with growing frequency, wearers of tracking technologies are relying heavily on nightly generated data to evaluate the quantity and quality of their sleep, and to make adjustments based on the data. Rather than waking and taking time to consider how we feel and how long we slept, we are making assumptions based upon data that may not be reliable. Once again, while the tracking is not, in and of itself, bad or negative, if it is used outside of self-assessment or real-time monitoring of actual experienced levels of tiredness or restedness, we forgo a strong and developed sense of really knowing ourselves. Consider, for example, research reported in the *Journal of Clinical Sleep Medicine*. Subjects were found to make inferences based upon fitness tracker data that caused them to self-diagnose sleep disturbances that were not clinically founded. In other words, our reliance on data can sometimes get us in trouble!

DEVICES YOUNG PARENTS USE AND THEIR EFFECTS

Other forms of digital monitoring provide similar forms of detachment from our embodied experience and assessment skills and deter the practice of self-discernment and assessment. The preponderance of technologies for new parents is of particular interest in this regard and highlights the growing push to outsource our ways of knowing and experiencing ourselves and others.

Of the many anxiety-provoking milestones in life, becoming a parent presents unique opportunities for delving into unknown territories. Anxious to provide excellent care for their children, new parents are faced with a steep learning curve at the same time they are generally deprived of sleep, social support, and general self-care. This sequela of realities often leaves parents feeling overwhelmed.

In states of change in general—and in states of exhaustion, of unknowing, and of high stakes in particular—humans look for certainties to hold onto. Developers of new technologies created for parents and children know this and now offer a plethora of products designed to take the "guesswork" out of parenting. Changing pads include digital scales and touch screens so that parents can record the number of diapers changed and send weight measurements to their smartphones at each change. Wearable monitors integrated into infant sleepwear record movement, skin temperature, respiration patterns, and body positions, feeding the data to smartphones as well.

Again, none of these new technologies is inherently "evil." Keeping track of how many minutes your children have slept, how much they have eaten, or how many diapers they have gone through offers parents the ability to confer with data when they need to assess the condition of their children or the quality of their own parenting response. In fact, this kind of data often provides a "comfort point" when one's emotions and nerves are frazzled. New parents in general, however, are often hyperaware of these measures of their child's "normalcy" or lack thereof. When a child sleeps through a feeding, drops a bit of weight, or has a fever, there is a tendency, in the early days of getting to know each other, to go into overdrive to find a way of resolving what seems abnormal. This is not a new phenomenon; parents for centuries have worked to balance the recorded data (e.g., number of diapers soiled, minutes slept) versus the nuanced observations of how a child seems to be doing on any given day. The accuracy and shareability of the data provided by digital devices, however, amps up the process significantly, putting today's parents particularly at risk of trusting their devices over their own observations of their children. Such overreliance on devices also enhances the notion that what we can know and learn based on our own careful observation or parental "gut" is not enough. Ultimately, it may even interfere with getting to know and understand our child or contribute to anxiety when we don't know how to evaluate or consider all the data to which we have access.

Additional outcomes of a parental focus on collection, assessment, and interpretation of Baby's "numbers" may include the underdevelopment of parental assessment skills and the shortchanging of the important parent-child bonding process in realistic and embodied ways. For secure attachment between a caregiver and a child, the keystones remain body-to-body, face-to-face, voice-to-voice, skin-to-skin interaction with rich sensory-motor activities. In other words, focusing on isolated data rather than on the entire being of the child can trigger a disruption in connectedness or attuned parenting.

I have noticed this with parents who use video monitors with their very young children. While auditory monitors let parents know when a child makes a sound, video-enabled devices allow parents to see whether the child is sleeping or awake. If parents assume that the time the child spends awake is not restful and repeatedly intervene with attempts to get the child to sleep, they may miss the reality that their child rests even in wakeful states. Their own interpretation, based on what they can see on a screen, forces them to foreclose on letting a child find his or her own way of being in the world. In sum, perhaps most important, reliance on devices often prevents parents from trusting their own inner wisdom and developing their own embodied relationship with their child—basically, in developing their "parental gut."

An antidote to this can be found in helping parents become less attentive to black-and-white, collectable, and recordable data and more attentive to the bigger picture. Does Baby seem to be thriving overall? Does she respond to attempts to comfort her? Is she filling out (rather than gaining weight with each feeding)? Is she bonding with parents in meaningful ways? When an infant is cared for by loving adults, this kind of quiet inventory is taken intuitively and, often, outside of conscious thought or recording. A baby cries and an attuned parent considers what the tears mean. Is it time for a feeding? A diaper change? Has there been too much or too little stimulation? Is it naptime? This kind of unconscious assessment requires that the parent tolerate the discomfort of the child (as evidenced by crying or other forms of communication) long enough to discern what type of intervention is most needed. This kind of nuanced relationship cannot be developed with data alone.

The movement toward depending on observable/measurable data to inform us of how we are doing extends well past infancy and is important for all of us. In the discussion above, there is an underlying assumption that if a caregiver is consistently present, emotionally connected, and observant, he or she will begin to develop a rhythm of discerning both subtle and obvious/measurable cues given by the baby to determine the baby's needs and preferences. A combination of knowledge including how many minutes Baby slept or spent contentedly resting (objectively measurable) plus important information about Baby's mood and temperament before and after the rest time (more nuanced and "felt" knowledge alongside "known" information) helps the caregiver understand what is needed for the child to be fully rested. The number of minutes alone does not give a full picture.

The same is true for all of us, regardless of our age. Our trackers can only tell us part of the story of how we are doing. They can only give us one type of data. For a full picture, we need an ability to assess and discern a full range of realities, including those that can be measured in objective ways and those that cannot. When we yawn, are we tired or bored? Do we need fresh air or a nap? Once we determine the nature of the cues our body gives us, it is up to us to choose the best and most fruitful path of meeting our needs . . . or not. Our devices cannot do this for us.

BABIES AND TECH

It is incredibly important for us to take a careful look at the seemingly innocuous and passive exposure of young children to technology. We, as a culture, have grown shockingly comfortable with children, even babies, interacting with technology. Although the venerable American Pediatric Association recommends limiting screen exposure during infancy and toddlerhood, popular teething toys with smartphones inserted into them make the screen the focus more than manipulation of the actual toy. Toddler toys that have historically offered imaginative opportunities for vocalization now make digitally produced sounds for the child, and those toys that once required kinesthetic movement by the user now have migrated much of the physical manipulation to screens. Instead of playing with makeup on their faces, children insert tablets into digitized mirror holders and apply digital makeup onto images of their face. Rather than crawling on the floor and pushing a toy car through an imagined roadway, children place wirelessly connected toy cars onto tablets that move the roadway under the merely rotated car. Such dynamics integrate screens as a "normal" and consistent part of a child's development, making the screens difficult to extricate or moderate later. Introducing devices at very early ages also threatens to inoculate children to high levels of stimulation, making less-exciting forms of embodied stimulation seem too dull to hold a child's attention.

The short- and long-term impacts of exposing very young children to digital devices are not benign. While peer-reviewed research may not have caught up to these effects, common sense tells us that once a person is used to a certain level and type of stimulation, it is difficult (although not impossible) to turn back. Also, it's easy to ascertain that exposing our young ones to technology rather than having them experience life through rich sensory-motor activities and loving attunement encourages being disembodied from the get-go. Thus, making sure that technology-based stimulation is balanced with consistent, warm, nurturing face-to-face and body-to-body relational connection and pure, child-directed play (when children don't have to respond to external stimulation but create their own play instead) is imperative for the health and well-being of the developing child.

DISEMBODIMENT

Being out of touch with our own body's ability to inform us of its needs, capabilities, and preferences leads us to live in what I call a "disembodied" manner. If we habitually switch off our alarm and look immediately to our sleep-tracking device before we ever consider how rested we feel, we are demonstrating a kind of disembodiment. When we reach for our phones whenever we find ourselves pausing (in line, at the end of the day, while we wait for a meeting or class to begin, when we wake up in the middle of the night), we miss out on opportunities to discern what our body and mind is and has been experiencing and what our body and mind might need in the moment.

The concept of disembodiment is important because we function best when we feel confident about caring for and regulating the body in which we live. Not only does being disembodied lead us to feel almost alien unto ourselves, but it can also lead to a variant of what psychologists refer to as "compensation." Compensation, considered a mature or higher order defense mechanism, is the counterbalancing of perceived or experienced weaknesses by emphasizing strengths in other areas. When we defer more to devices than to physiological impulses or needs, we begin to feel less competent in developing patterns that attend to our actual kinesthetic and sensual needs and preferences. This, in turn, causes us to increase our competence and interactions with devices to perform the needed discernment. Simultaneously, the skills we need to be healthy are not being practiced and our ability to use those skills is weakening.

As an example, consider the way we rely on our devices as "decompression tools" at the end of long days. Rather than bring conscious awareness to a felt experience of tiredness and offer the body meaningful forms of rest, we turn automatically to our screens for distraction. We do the same with boredom. Because our devices are so easily accessed, require relatively no preparation or cleanup time, and can be engaged alone and wherever we may find ourselves, it's no wonder that's where we now direct needs that were once experienced and addressed in the body.

Even our sense of place in the world is impacted by our reliance on our devices. I notice this when I travel. I can currently visit a new city and never need to truly know (or notice) where I am. By relying on digital directions to get me from the airport to point B, I simply keep my eyes and ears tuned to my device rather than gain any kind of contextual understanding of the new place that I am in. While the same app that is providing me with turn-by-turn directions could easily be used to give me a wider picture, I rarely use it in that way because I never have to. I think many of us do this much of the time. We don't rely on our own directional capabilities, we don't look up and around, and thus we don't truly inhabit the actual space where we are. Add to this the tendency to search for the familiar and well-reviewed, and it's easy to see how we miss out on new sights, sounds, smells, and even tastes. Our losses!

EMBODIMENT AND SEX/SEXTING

No conversation about embodiment can be complete without addressing the way our devices allow for disembodied sexual exploration and expression. No longer limited to in-person or voice-to-voice communication, people can now engage in sexually intimate relationships from a distance and in asynchronous ways. Turning our sensual bodies into one-dimensional images or our words into messages sent with no regard for the context or timing in which they are received means we can now experience increasing amounts of sexual exploration and expression at a greater distance from the physical presence of others. Even if we experience a sense of stimulation in these encounters, any direct physical action remains "self" with "self," and this separates us from the complex, beautiful, tricky messiness of in-person embodied connection and intimacy.

Perhaps the most commonplace example of disembodied sexual expression is sexting. "Sexting" refers to the sending and/or receiving of sexually explicit photos or messages via digital devices. In the early days of texting, sexting was seen, almost unanimously, as a deviant and overly risky behavior. As time has gone on, however—and as sexting has become mainstream—the picture isn't so clear. Some research links sexting with higher chances of risky sexual behavior, but other research finds sexting to be a relatively healthy form of sexual communication and exploration.

In a groundbreaking exposé, *Atlantic* reporter Hanna Rosin uncovered the real complexity inherent in the current sexting trends. Rosin covered the story of a small rural town in Virginia where students, parents, and law enforcement officers found themselves dealing with differing expectations and norms regarding what is and is not "appropriate" to text. While the adolescents who sent naked photos of themselves felt uncomfortable about being "caught," when law enforcement officers were tipped off to Instagram accounts where a plethora of sexted photos were catalogued, they also didn't succumb to the kind of fear and panic their parents did. Perplexed by the new norms around communication, parents and law enforcement officials found themselves divided on whether the individuals sending the photos or those reposting them should be disciplined, or if they should treat the epidemic as inspiration to do some new training and talking with the teens in town. They figured that cracking down too hard might have serious consequences on the senders' or the posters' legal records but that under-responding could cause unknown harm.

This dilemma brilliantly illuminates the complexity of sexting. As a behavior and communication form, it is here—and it is here to stay. Whether sexting is healthy or harmful depends much on the context of the relationship of the people involved, their ages, the trustworthiness of both, and, most importantly, forethought, insight, and consent. It's also important to note that nudity and the appropriateness of sexual language in general mean vastly different things to different people. Among today's youth and young adults, body "positivity" (a valuing and affirming of all shapes and sizes of bodies) brings with it a relative casual consideration of nudity. Similarly, what one generation sees as explicit sexual and/or vulgar language may not be considered such by another. Even within the same age group, some images and words might be viewed as nonsexual by senders and posters but be *received* by others as highly sexualized. These very contradictions make a conversation about sexting and sexual exploration in digital spaces challenging, to say the least.

In general, it's difficult for me to say that sexting leads us to be more healthfully embodied as people. While it may provide a way for us to promote or stimulate our sexual impulses and our physical bodies, the fact that it happens divorced from the complexity of in-person interaction means that it inherently keeps us in a self-focused loop. It also separates us from the in-the-moment practice of being with another person who is also in his or her own skin. Whether these dynamics of "separateness" from the body and communication are helpful or not depends on many complex situational elements. While sexting may be healthy for part-ners within committed relationships or between consenting adults who have weighed possible risks and costs, it might be unhealthy and even destructive to a developing individual's ability to discern and carry out metered expressions of intimacy and sexuality within embodied relation-ships. Especially when the sexting happens during times of impulsivity or in response to feeling coerced, the effects are likely to be very harmful. Given the fact that much of the sexting that happens between teens and tweens fits into this category, it is imperative that parents talk with their children long before they think they need to about sexuality, expressions around sex, how to discern the trustworthiness of others, how to pause and consider potential consequences (even or especially when feeling sexually and emotionally stimulated), and how to communicate about consent. Even if these conversations are uncomfortable, it is far better that they come before sexting starts so that the child knows he or she can come to the parent/adult if things go sideways. Again, these conversations are crucial—because things often do go sideways in the arena of sexting.

Part of the risk is related to the fact that sexting is a disembodied form of communication. When used as a way of substituting for person-to-person contact or communication, we need to be clear about the goal and to discern whether it is healthy. If we use sexting to advance a relation-ship sexually beyond what the emotional quotient of the relationship can handle, we must ask ourselves if this is in our best interest. If sexting is done impulsively, without forethought regarding the trustworthiness of the recipient or the "unprotectable" nature of the internet, this, too, must be addressed. Once tapping that little arrow, the sender is completely out of control—and much can happen with a photo or words in online spaces! Revenge porn, or the reposting of sexual content to hurt or shame someone, is much more common than many imagine, with 10.4 million Americans being the victim or threatened victim of it. Victim blaming after revenge porn posts appear is particularly harsh and must be avoided. This is just one more place where the complexity of the issue requires intentional forethought and discussion.

TAKE ACTION

Ideas for Parents: Talking about Sexting

DON'T WAIT TO HAVE "THE CONVERSATION."

Talking with children and adolescents about sex is difficult for many parents. It's time, however, to get over this discomfort and do the hard and crucial work of having the first conversations about all things related to sex and sexuality. Culture is more than ready to do so, and it is comforting to children to know that the grown people in their worlds understand and welcome all the realities that go along with having a body, including being a sexual being. Late elementary school–age children will likely encounter sexting and porn sooner than most adults imagine. Helping them be prepared can go a long way toward how they handle the challenges related to both.

USE ALL THE LANGUAGES.

While teaching children anatomically correct language and being comfortable with using these words is important, it's also imperative to help them navigate the landscape of pop culture language and symbol use. In today's typed communication reality, kids often send messages that mean one thing to them and another to the recipient. Find sources that can help you learn all the languages your child needs to master. This includes understanding jargon related to texting and the connotations that go along with emojis.

HAVE BODY-POSITIVE, NON-SHAMING CONVERSATIONS.

It's easy to let anxiety, fear, or anger drive conversations about sex in general and sexting in particular. Our children need us to be able to regulate our own emotions in such a way that we can make space for theirs. Approach conversations about sexting with statements such as, "As you text with friends, you are likely to receive some that make you feel all sorts of ways. Words and pictures might be sent that make you feel sort of excited, also weird. Some of these might include naked photos or comments related to sex. It's pretty normal to receive these; it's also normal to feel unsure about what to do about them." Such caring language can be a huge help in keeping the conversations going. Basically, children need to know that their bodies are wonderful, that it makes sense to feel proud of them, and that it is important to thoroughly think through what might happen if they share naked or provocative images of themselves. They need to feel that adults understand their sexual impulses and desires for exploration, that we acknowledge the "normalcy" of enticing online sexuality, and that we want to help them navigate this reality in their lives. We also want them to know we are not afraid of these realities and will not overreact if they find themselves in a bind. We want them to come to us even if they've made a misstep—especially if they've made a misstep. We want to be their loving resource.

HELP CHILDREN UNDERSTAND THE IMPULSIVITY WITH WHICH HUMANS RESPOND.

While they may feel inclined to overlook potential consequences of sharing sexts or may believe they can trust those they send them to, help them learn to pause in response to the many forms of impulsive action online. Given that many early sexting experiences happen in the middle of the night with children who are sleep deprived and at developmental periods when self-control is not yet established, helping them anticipate and preset a habit of pausing in response to "temptation" is a huge gift to children. Literally, helping them make a plan can be a lifeline. Say things such as, "Let's pretend it's the middle of the night and you have your phone. Friends who are having a sleepover begin sending you photos of themselves with little or no clothes on and dare you to do the same. You're feeling pretty excited that they chose you to send messages to and are also excited about how they might respond. You don't want to seem like a loser. What are some ideas of how to act in this situation?" Then provide some helpful ideas to seal the deal.

WATCH HOW YOU SPEAK ABOUT OTHER CHILDREN/ADULTS.

Practice nonjudgmental awareness. Our kids are watching us. When they hear us put other people down for behaviors they themselves may have engaged in or been tempted to engage in, our kids get the clear message that we will put them down as well. More than ever, children need parents who will help them navigate. They need to know that parents and other caring adults will be able to handle their own feelings well enough to help them deal with the unbelievable and never-before-navigated waters of life in this time. For them to believe they can come to you when they have made a mistake, they must know you will be able to tolerate the discomfort without becoming dysregulated or shaming them.

FIND SOMEONE SAFE TO TALK WITH SO YOU CAN DO THE ABOVE.

None of this is easy. The easy options are to put our heads in the sand and to make unrealistic demands on our children to simply resist and obey. When we have places where we can be supported and cared for as we ourselves navigate these murky waters, we will be much more able to suspend our own reactivity in order to educate and nurture our children through approaches and missteps to sexual exploration on- and offline. Resist the temptation to believe that everyone else's children are perfect and have never struggled! Instead, find those who can share authentically with you and who will support you as you, in turn, support your child.

MOTIVATION AND SELF-SOOTHING

Less aware of our own sensations and physical realities than of our devices, we have come to live comfortably in various levels of disembodiment. By checking in more consistently with things outside ourselves than with our own emotional, intellectual, and physiological states, our knowledge of our internal realities suffers. It's not so much that we don't want to pay attention to our hunger or our restlessness; it's simply that we don't notice these realities due to the enticing pull of the digital spaces in which we are invested. This often leads to displacement when it comes to meeting our physical needs. For example, when we're tired, instead of going to sleep, we pull our laptop onto the bed with us and binge on a new show. Or when we're anxious while waiting, instead of accepting the stickiness of the situation and working through it, we check social networks or play a game on our phone as a distraction. In other words, we sidestep addressing our feelings and needs by reaching into digital spaces. This leads us to experience varying levels of disembodiment. Life lived this way is costly.

One serious side effect of our growing disembodiment is a sense of incompetence surrounding our ability to stimulate or soothe ourselves both physiologically and emotionally. When we invest so much of ourselves in external spaces, we slip into relying on those spaces for comfort and motivation. Every time we do this, however, we subtly reinforce a cycle that makes us disinclined to develop our own meaningful and effective self-soothing and self-stimulating skills. When we rely on digital devices to motivate us, we miss opportunities to hone an internal process by which we spur ourselves on to approach and master new tasks. Similarly, when we retreat into the expansive digital opportunities for distraction and engagement to soothe us when we are anxious or stimulate us when we are bored, we fail to cultivate other important life skills.

Truly a vicious cycle, this reality makes us tend to defer to devices for stimulation and soothing on a regular basis. When we haven't practiced calming our minds and bodies without the help of technology, we feel less capable of doing so on our own without digital assistance. When the constant distraction and numbing of an always-changing social network feed, streaming queue, or video game make us feel less actively upset and stirred up, it's easy to believe we are physiologically and emotionally soothed—and we leave it at that. The same goes for stimulation. With access to never-ending visual and auditory stimulation, it's easy to bypass any effort to motivate ourselves to action completely on our own.

Another consequence of reliance on devices for self-soothing and stimulation is that we are left bereft of the sense of confidence that internalized, mastered skills impart.

In other words, the dependency on technology for emotional and physical regulation comes with a cost to our very sense of capability in the world and in our ability to regulate ourselves. The side effects of this are not pretty. Whenever we as humans lack a skill important to our survival, we experience an increased chance of anxiety. So, with underdeveloped abilities to cope with waning motivation in the face of life's challenges or to calm ourselves when emotionally overwhelmed, we live with a constant sense that we just don't have what it takes! This subtle, unconscious awareness further cements the cycle of turning to our devices rather than developing our own capabilities.

One of the areas where this difficult reality is evident across all genders is pornography, which we'll talk about in more depth in the next chapter. For our purposes here, consider that when people have habituated to channeling boredom, agitation, and sexual energy toward engagement with online porn, they most likely have centralized functions related to both stimulation and soothing in the platform itself. Engaging porn in ways reflective of other types of online engagement, these individuals switch between multiple sites in rapid succession in order to experience higher and higher levels of stimulation. Uncertain of how to soothe themselves in the face of embodied challenges, these same individuals often rely on the release of orgasm to create a sense of calm or soothing. Over time, this cycle of stimulation followed by a perceived sense of release and calm becomes so reinforced that the individual feels uncertain of how to achieve calm *without* the precursor of high levels of visual and manual sexual stimulation. Making matters worse, because the mind and body are so interconnected, the cycle take its toll on the physical body: Porn users are usually left with a diminished ability to maintain arousal in embodied-life situations with consensual, present partners; specifically, men who use porn often cannot maintain an erection with their embodied partners. When this occurs, sex and sexuality are often accompanied by a sense of anxiety and fear. This is especially difficult for emerging young adults, who may have no other information about what embodied sex might really be like and who believe they will never be able to achieve the kind of sexual performance they see on screen. In addition, porn users may feel they cannot get help, fearing the shaming and lack of understanding that will result from disclosing their experiences.

Whether in relation to porn use or any other kind of engagement, stepping away from the habituated tendency to turn toward our devices requires some degree of insight and self-discipline as well as being able to self-soothe. Self-soothing requires an ability to get centered and assess our emotions and physical sensations. It requires a turning inward of sorts that allows us to focus on what we need to regulate ourself apart from anything external. But that's easier said than done. Someone who habitually uses technology for self-soothing will very likely experience anxiety when he or she attempts to put down the device to determine what's truly needed to self-regulate. This anxiety fuels the cycle of dependency, leading the person to believe that distraction (via device) is easier, and therefore better, than internal self-assessment and attempts at regulation apart from technology. That is, when we try to step back from our habits, our physiological and emotional sense of fear will be triggered. To break the cycle and reinvade the life within our own skin, we must be able to tolerate the anxiety and to soothe it. Only then can we begin to embrace a gentle curiosity about what our body needs and to live from a confidence about who we are as embodied beings.

THE CONNECTION BETWEEN DISEMBODIMENT
AND BODY DYSMORPHIA

"Body dysmorphia" refers to a fixation on imaginary or inflated defects in one's appearance. In a world where hundreds of edited images confront us daily, it's difficult to remain content with a body that is human and, therefore, flawed. This is an issue for all genders. Hypermasculine and muscular physiques appear alongside unhealthily thin ones on our screens. While most of us know *consciously* that these images are highly edited, the sheer mass of these airbrushed and Photoshopped images impacts our *unconscious* ability to feel content with our own flawed bodies. This is especially true for children, adolescents, and emerging young adults, as well as adults who suffer from feelings of inferiority. We curate our online profiles to try to amplify our "positives," but we can never seem to attain the kind of perfection the images all around us project. Self-improvement apps, created to edit and filter images prior to posting, are frequently downloaded in an effort to make us appear more attractive, but they end up putting us in a bind as we realize we can't live up to our own images when people meet us in person!

It is up to us to boldly shape our communal sense of beauty and attractiveness—and to do so with a keen eye to authenticity and attention to the "insides" of those around us. It takes deliberate discussion and consistent reminders to keep ourselves and those around us from believing the lies our devices tell us about what is beautiful and healthy and authentic. It is up to all of us to keep these conversations going, to be aware of the example we set with our own self-talk and the projection of images of ourselves, and to esteem those around us for both their character and the beauty of their insides and outsides in real and truthful ways.

TAKE ACTION

Practicing Living an Embodied Life

- Work to love your body with all its imperfections. Identify things about your body that you appreciate or enjoy. Practice viewing your shortcomings with graciousness and then redirecting your attention to a trait you appreciate. Thank your body for the things it does well.
- Learn to listen to, soothe, and—with warmth and gentleness—care for your body.

- Experience your sexuality and desires in safe ways that are respectful of yourself and others.
- Cultivate your intuitive intra- and interpersonal senses.
- Cultivate your physical senses (see the next section).

CULTIVATING YOUR PHYSICAL SENSES

One important way to develop and harness the necessary self-soothing skills that will enable the process of living life in real embodiment is to attend to all our senses each day. Unfortunately, when we over-rely on technology, we limit ourselves to visual and auditory stimulation, with some eye-hand coordination practice thrown in for good measure. Our other senses often go neglected. Even still, our miraculous bodies are constantly attending to what our senses are taking in at some level or other and adjusting appropriately. We can either be aware of this or it can happen in primarily unconscious ways.

If we *proactively* prepare ourselves to be aware of our interactions with the embodied world and similarly prepare our environments to be "settling" (pleasant, appropriately challenging, or tailored to meet our personal needs) and appropriately stimulating for our senses, we can truly live in rich, embodied ways. This intentionality can be called being "mindful" and requires hard work and dedication. We are all prone to slip into the easiest, most readily available, and most familiar ways to stimulate our senses, which mean gravitating to technology to see, hear, and feel things. But we can do better. The payoff can be a richer life with a greater sense of awareness and a larger pool of opportunities for stimulation and soothing. Here are some simple ideas for stimulating each of the sensory fields:

Sound

- **Cherish silence.** Maintain an ability to be in and with silence by creating it. Leave the television off for a while. Choose specific times to drive with no radio/digital content. Walk/run/work without earbuds. Let a podcast or two go unheard. Visit places such as libraries and empty places of worship, where silence is the norm. Sit for at least ten minutes and pay attention to the sound of silence. What do you hear? What do you feel? What happens with your other senses as the need for active listening falls away?

- **Turn down the volume.** Set the volume of your laptop/television/ phone a bit lower than normal. Notice how this feels. How does working at listening feel?
- **Take earbud breaks.** Set aside times when everyone in your home or work setting is earbud free—and maybe even device free for a while. This lets everyone in on what everyone else is doing and makes you aware of how much stimulation is influencing each person in your environment.
- **Vary your playlist.** Listen to a variety of genres of music/content. Challenge yourself to stretch into new styles. To find compelling elements in what you hear, listen past the point when the newness bothers you or your distaste falls away. Listen to an entire recording as presented. The artist ordered it for a reason. Notice how this feels.
- **Try going lyric free.** If you must keep your earbuds employed, listen to lyric-free music when trying to study or work. Experiment with genres such as baroque, world, jazz, or electronic. What do you notice as different themes within the music emerge? Which forms increase your attention to the tasks you are working on? Which distract you?

Vision/Sight

- **Declutter.** Pay attention to "visual clutter" in your home, work, or school environment. Notice how it feels to look at the cluttered spaces versus spaces free of visual stimulation or clutter. Regardless of your style or temperament, we all need quiet places for our eyes to rest. Ensure that you have places in your home/work space/classroom for your eyes to land with little to no stimulation. Practice drawing your attention to these spaces when you are overwhelmed or need a break. Notice how it feels for your eyes to have a place to rest.
- **Renew your view.** Give yourself new things to look at periodically. Take a new way home, visit a place you've never been (even just a new neighborhood in your city), or take a hike in an unfamiliar setting. Switch out or rearrange the art in your home or office. Pick up a children's picture book or a photographic/illustrated coffee-table book, and set aside time to take it in at a slow pace. Notice what draws your eye and what repels it.
- **Find eye feasts and indulge.** Provide yourself with visual complexity. Art museums, image-rich magazines and journals, pattern-based coloring books, and natural settings with a variety of foliage all stimulate the visual field in important ways. Make sure that screens aren't your only source of visual stimulation.
- **Think about lighting.** Light impacts our sense of visual comfort versus discomfort. As a general principle, overhead lighting (that is not

highly designed or managed) is hard on the eye and creates shadows on the faces of those with whom we interact. Lighting at face level (e.g., table lamps that are right at face level) is easier on the eye and provides a more comforting environment. Make some changes according to these guidelines and see how the changes make you feel. Notice how the light changes in your environment as the sun goes down, and try to make the change from natural light to artificial forms of light more seamless.

• **Power down prior to bedtime.** Digital devices emit powerful doses of light that stimulate neurotransmitters and hormones related to wakefulness and stimulation. Try powering down all electronics at least thirty minutes before trying to sleep. Increase this to sixty or even ninety minutes over time. Over the course of a two-week trial, notice how your sense of restfulness waxes and wanes as you eliminate screens closer to bedtime and when you are in bed.

Smell

• **Pay attention to scents.** Notice how naturally occurring smells in your daily life impact your sense of awareness and attention. Are there fragrances in your home/office/classroom that distract you or overly direct your attention? Are there any that enhance? Try plugging your nose when you take a bite of food. How does the lack of olfactory stimulation affect your awareness of the texture and taste of your food?

• **Try essential oils.** Olfactory stimulation is often encountered by chance rather than by intention. We tend to notice smells when they occur naturally, and we encode them with emotion in our memories. It can be powerful to use olfactory stimulation by design—engaging fragrances to stimulate or soothe, to heighten awareness or to set a mood. To do so, use them on pulse points on the body in essential oil form or throughout your living/work/learning spaces with infusers or candles. As a very general rule, citrus scents (lemon, lime, orange) invigorate and stimulate, while plant-based scents (rosemary, clary sage, eucalyptus) soothe and relax.

• **Go international.** Go to an international market or restaurant. When and as you can, close your eyes and focus only on the smells. How do these new smells make you feel? What do you become aware of?

• **Grow fragrant plants.** Experiment with growing a fragrant plant where it can be easily accessed. Rosemary and lavender are relatively easy to grow. Once the plant is mature enough, break off a small piece and rub it between your fingers. Take the smell in as you breathe deeply to create a sense of calm. Work to actively link the fragrance with the embodied experience of feeling calm.

Taste

- **Spice it up.** We often gravitate toward tastes we know and with which we are comfortable. Periodically stretch yourself to try new flavors and textures. Do this in small and manageable ways. Try a new spice. Buy a small bag of uniquely flavored potato chips or an unusual (to you) piece of candy at an international market. If you naturally gravitate toward sweets, try something savory or vice versa. This can be done with drinks such as tea as well as with food. If you have access to a good tea shop, stop in and try a smoky blend. Notice how you anticipate and then taste the flavor.
- **Go bland.** Try food that has not been flavored or seasoned. If you drink coffee or tea with sweeteners, try the drink without. If you are used to processed foods, seek out a meal or food experience that is preservative and enhanced-flavor free. Notice the differences, even if you don't prefer them.

Touch

- **Mix it up.** Whether we are consciously aware of it or not, our skin is constantly perceiving what it touches or is touched by. "Waking up" this perception can lead to greater sensory awareness. Provide yourself opportunities to feel things that are rough, smooth, wet, dry, hot, cold, and more. Pay real and focused attention to how they feel and what kinds of sensation you experience as a result.
- **Add touch to learning.** Some individuals can increase their focus and attention in life by simply having something to touch or play with during learning experiences. For these people, knitting or crocheting, a handful of Silly Putty, a bowl of Kinetic Sand, a small scrap of carpet or AstroTurf, or a rock might become important tools for maintaining focus and attention. Individuals who are kinesthetically/body smart benefit immensely from attending to the body in this way. When they do not actively work at getting the kinesthetic/physiological stimulation they need, they are at risk of using substances and people outside themselves to stimulate them. Drug and alcohol use, sexual acting out, and self-injury can become serious issues for these individuals.
- **Experiment with weight and swaddling.** Sometimes our bodies can benefit from feeling "contained." If we don't have others to hug or hold us, we can wrap a blanket around ourselves and pull it snugly. Warmed, rice-filled compresses also can be used over closed eyes or the chest to create a sensation of calming. Therapeutic weighted lap pads and blankets are also available for sale in a variety of stores and can be found online through a Google search.

4

Our Bodies and Brains on Tech

Our interactions with devices aren't just changing the way we are *in* our bodies; they are creating *actual changes* to our bodies and brains as well. Given the manner in which our physical selves adapt to and are altered by the circumstances around us and the nearly constant presence of devices in our environs, there is no way we will not be impacted by technology. While some changes may turn out to be positive ones, some may not. The good news is that our bodies and brains are dynamic systems, able to repair and replenish themselves. If we educate ourselves about how our interactions with screens might impact our very physiology, we may be motivated to moderate our use in such a way as to maintain optimal health.

Perhaps one of the most common associations drawn between screen engagement and physiology is the link between heavy screen use, sedentary lifestyles, and obesity. Cries of sitting as a public health issue have cropped up everywhere. In response, gaming companies have invented more "active" technologies, promising to reverse the health threats by getting people moving as they interact with their screens. Think gaming systems with hand-held controllers allowing people to bowl as though they were in a bowling alley and dance on mats to onscreen choreography as though they were on a stage or dance floor. In fact, some research shows promising health effects from active video games. Shortly after their release, however, those same games were the subject of the popular press revealing this reality: many players find ways of playing even active games from a sitting position!

While correlation does not infer causation, it does make sense that when we spend more time in front of our screens, we have less time to be physically active. Twenty-five years ago, research out of Harvard found a link between the sedentary activity of television viewing and obesity. Since then, research has continued to show that a high amount of screen engagement, in the form of both television and smaller-screened devices, is related to obesity in adolescents and adults. A recent Harvard study expounds on this relationship, noting that of the nearly twenty-five thousand American teen participants, the 20 percent who spent more than five hours per day in front of screens not only were twice as likely to consume the sugar-filled drinks advertised online but also did not get adequate sleep or exercise. In turn, these adolescents were 43 percent more likely to be obese than the participants who spent less time with their devices.

It isn't just the exposure to increased advertisements, the lack of sleep, or a sedentary lifestyle that puts our bodies at risk with increased screen time. As we discussed in the last chapter, less time spent *in* our bodies means less practice. When we have less practical experience moving our bodies, we become less familiar and comfortable with doing so. As we exert ourselves less, our bodies lose resilience, strength, flexibility, and sometimes even capabilities. The truth is that our bodies require movement to stay healthy, and our engagement with devices often serves as an impediment to such movement.

According to Dr. James Levine, director of the Mayo Clinic/Arizona State University Obesity Solutions Initiative and inventor of the Walk-Desk, "Sitting is more dangerous than smoking, kills more people than HIV, and is more treacherous than parachuting. We are sitting ourselves to death." Given that a growing number of today's hobbies encourage excessive sitting, staying physically active is a real problem. Gaming is one such hobby. In 2013, gamers reported playing, on average, six hours a day. Now we have Twitch, a site that offers twenty-four-hour live streaming of gamers playing in real time. Using this popular app, gamers work to establish large followings by streaming for long periods of time, often raising money for charities while doing so. In 2017, a gamer walked away from his game camera at hour twenty-two of what was to be a twenty-four-hour gaming session and never returned. He died while off camera. Another gamer who streams to 579,000 followers on Twitch has reported high cholesterol, exhaustion, and heart problems related to his profession. Further, in a popular essay, "Dying to Stream," another gamer, with forty thousand followers, claims that his heart surgery was the result of his spending seven to eight hours per day gaming. Following the death of one of his fellow gamers, he took stock of the effect of gaming on his physical health and found himself with 100 percent blockage in an anterior artery surrounding his heart and a 75 percent blockage at the major three-way cardiac junction affectionately called "The Widowmaker." These may be extreme examples. Nonetheless, they serve as a dire warning.

TAKE ACTION

- Take breaks from screens for movement throughout the day to help you stay not only healthy but also engaged.
- Get into the habit of walking away from your devices at least every hour to get fresh air and move both your legs and small muscle groups. Even stepping outside for three deep breaths can make a difference.
- Try many different types of physical movement. Doing so will help you stay flexible both in your physiology as well as in your beliefs about your body's capabilities.
- Associate one of your tech hobbies with a set of basic and easy-to-do-wherever-you-are stretches. Do these every time you engage that tech habit. For example, do a sun salutation or two every time you pick up your game controller or log on to social media.

OUR POSTURE ON TECH

Other risks associated with near-constant engagement with devices involve postural changes and potentially related pain. A joint research study by Harvard School of Public Health, Brigham and Women's Hospital, and Microsoft showed that holding a device (a tablet, in this particular study) low in one's lap places the neck muscles as well as the small, interlocking bones at the top of the spine (the cervical vertebrae) in unnatural postural alignment. The resulting effects range from strain in the muscles and nerves to overexertion in tendons, ligaments, and spinal discs. Pain, as well as possible damage, can result from this type of strain.

Specifically, more and more medical practitioners are starting to diagnose patients with what they are calling "cashew" posture. We have all seen people sitting down texting, with their head slumped forward, hanging in front of their spines. The neck and upper back muscles experience a constant strain in this position, and chronically contracted muscles can become oxygen-starved and develop a buildup of toxic waste products due to lack of circulation. Technically called "forward-head posture," this malady is associated with headaches, neck and shoulder pain, muscle spasms, excessive upper back curve (thoracic kyphosis), nerve compression, and fatigue. With our regular screen use, getting into a repetitive cycle of poor posture and pain can be all too easy. As we age, these symptoms can get even worse!

Postural concerns don't only involve the back and neck either. How we hold our hands at our keyboards and our thumbs on our small screens also impacts our health. For example, carpal tunnel syndrome is not uncommon among programmers, gamers, and writers who use keyboards repetitively for many hours each day. In addition, thumb arthritis, historically a concern only for those of advanced age, is on the rise in children and young adults who rely upon their thumbs for typing texts on small screens.

TAKE ACTION

To prevent negative postural effects while using screens:
- Remember to step away from your devices regularly.
- Practice good ergonomics.
- Stretch regularly.
- Engage in flexibility exercises.
- Make sure your screens are level with your eyes when looking straight ahead.
- When using a keyboard, keep your back straight and your arms parallel to the floor and close in at your sides. Also, rotate your wrists occasionally.
- When using small devices, be sure to stand and stretch, shift your weight, and rotate your thumbs and wrists occasionally. Look up and around and intentionally stretch the top of your head toward the sky.
- When using any device, be careful not to round your shoulders or lean your head excessively forward.
- Practice mindful, thoughtful device engagement.

Even when we are not interacting with screens, we should be practicing good, healthy posture. To do so, imagine a string attached to the top of your head, pulling gently upward. Drop your shoulders back and tuck your pelvis under, letting your gaze fall directly ahead of you. Feel the soles of your feet pressed firmly into the ground. Memorize this feeling, and refer back to it throughout the day.

OUR EYES ON TECH

We also need to be vigilant about protecting our eyes when using technology. This discussion must start with an understanding of the impact of blue light, which is a color in the light spectrum that can be seen by the human eye. Unlike other colors of the visible light spectrum, blue light has a very short wavelength. Scientists tell us that the shorter the wavelength, the higher the energy, so blue light produces a high amount of energy. These short, high-energy blue wavelengths of light collide with molecules in the atmosphere, causing blue light to scatter everywhere. Historically, we've only had exposure to blue light from the sun.

On a positive note, blue light helps boost alertness, heighten reaction times, elevate moods, and increase the feeling of well-being. Because blue light's wavelengths are short, however, they tend to flicker more easily than longer, weaker wavelengths. This flickering creates a glare that can reduce visual contrast and affect sharpness and clarity, resulting in eyestrain, headaches, and physical and mental fatigue. The human eye's natural filters do not provide sufficient protection against blue light, and studies show that over time, overexposure to blue light can cause serious long-term eye problems—from retinal damage to macular degeneration to vision loss. That's why doctors warn us to wear high-quality sunglasses outside to protect our vision.

What does this have to do with digital devices? In the past ten years or so, an artificial form of blue light has become the mainstay of energy-efficient fluorescent bulbs, light-emitting diodes (LEDs), and the backlight of the smartphones, tablets, laptops, and televisions with which we interact every day. This massive application of blue light means we have exposure to it in ways and amounts not possible a decade ago. If we struggle to protect our eyes from the sun's naturally occurring blue light, consider how much more challenging it can be to add protecting our eyes from our nearly always-on computer screens and light sources.

While many apps and devices offer filters to help with blue light exposure, some consider these akin to filters on cigarettes. They are helpful but cannot remove all risk. For some, the use of blue light filters can lead to a false sense of security, enabling people to overexpose themselves by not taking the kinds of breaks that can be helpful to the eyes.

TAKE ACTION

To minimize the negative impact of blue light related to screen use, take the following steps:

- Take breaks from screens throughout the day.
- Make sure screens are not placed in front of windows, forcing your eyes to adjust to both light sources.
- Use lighting at eye level rather than overhead when working with screens indoors.

Before leaving this discussion about blue light, we need to address another of its impacts related to technology. Scientists tell us that the difference in using light as "illumination" (for our screens) rather than as "reflection" (as on paper and other objects our eyes take in) makes a difference not only to our eye health but also to our moods. The production and moderation of the neurotransmitters related to mood regulation and depression—dopamine, serotonin, and norepinephrine—are related to the solar day's natural patterns of light and dark. Exposed on a regular basis to new forms of light brought to us via screens, our very emotional states are impacted. Light also plays a huge role in regulating our sleep and the circadian rhythms by which our internal clocks run. Recall that I noted earlier that blue light boosts alertness—not something we want when it's time to go to sleep!

Participants in a study that compared reading on illuminated screens before bed with reading by reflected light (a paper book) before bed found that those reading from a screen took longer to fall asleep, experienced less sleepiness in the evening, demonstrated a reduction in melatonin secretion (which impacted the timing of their circadian rhythm/internal clock), and experienced reduced levels of alertness than those reading a printed book. These results show that nighttime exposure to e-books (and possibly all screens) pushes back the circadian clock and acutely suppresses melatonin. These findings tell us that screens in bed, or interacted with near bedtime, have an impact on sleep, performance, health, and safety. A preponderance of research replicates these findings.

TAKE ACTION

- In the evening, turn away from screens at least thirty, optimally ninety, minutes before heading to bed.
- Read one book in paper form for every book you read on an electronic reader device.

OUR EARS ON TECH

Just as digital devices pose threats to eye health, earbuds have the potential to impact our hearing. Noise-induced hearing loss can occur when earbuds are consistently set at high volume levels. If a person sitting an arm's length from someone using earbuds can hear what is being broadcast through them, the volume is likely at or near eighty-five decibels and can, if prolonged, cause noise-induced hearing loss. This type of hearing loss is irreversible.

TAKE ACTION

The National Institute on Deafness and Other Communication Disorders (NIDCD), part of the National Institutes of Health (NIH), suggests the following:

- Listen to earbuds at lower volume levels.
- Limit the volume level to a maximum of two-thirds of earbud capacity.
- Move sounds farther from the ear at times.
- Use ear protection when exposed to loud sounds to limit the chances of damage to the sensitive structures of the inner ear.

OUR BRAINS ON TECH

The brain is made up of complex electrical circuitry. Information comes toward the brain from one of the senses, which initiates a series of electrical impulses or an energy flow that creates what is called a "neural pathway," taking the information to places in the brain where it can be encoded or stored. This action, plus the release of chemicals (neurotransmitters) in the brain, leads to integrative functions that allow us to think, feel, and act. Neuroscience research tells us that the brain wires together where it fires together. This means that the structure of the brain is largely dependent upon situations that engage it—or get it "fired up"—to make learning and encoding happen.

When one of our senses (sight, sound, touch, etc.) presents information to the brain, that input triggers a series of impulses along (1) pathways that are already established or (2) those that are newly created. If we expose ourselves only to very mundane and repetitive stimuli, we will have fewer pathways and less complexity in the wiring of our brains. On the other hand, if we expose ourselves to a variety of novel and meaningful situations in optimal settings, our brains will create complex circuitry. This circuitry exists within a given structure that involves both gray matter (which serves as the power, or engine, of the brain) and myelin (which helps with efficiency). In a healthy brain, the circuitry, chemicals, and structure all work together to create an integrated system that is flexible and adaptive. Given how complex the brain is and how significant the factors that contribute to its health, it behooves all of us to understand how to maintain brain health.

Contrary to historical notions, the brain is not akin to a filing cabinet. It doesn't happen that we need a math skill, so our brain accesses the math file to help us. Instead, many regions of the brain work together to help us solve any given problem, and many factors contribute to the way in which we encode information as well as the ways we store and retrieve it. For instance, information presented to the brain along with high emotion is encoded differently than information that does not capture our emotional attention in any way.

Brain research tells us that brain health is largely dependent on both our physiological health and the nature and type of experiences to which we expose ourselves. Consider neurological development: When we are babies, our brain develops from the back to the front and from the bottom to the top. In the most basic description possible, we would say that the brain stem develops first and is fully formed at birth. This is the "reptilian," or "lizard," part of the brain, responsible for automatic functions such as respiration and heartbeat. This region also is highly involved with our response to threat and moderates our tendencies toward fight, flight, freeze, or faint reactions. Our lizard brain works in concert with the limbic system, which develops next and is partially formed at birth. The limbic system includes many structures of the brain responsible for motivation, drive, reward, and emotion. When this system does not develop in an integrated way, real difficulties in emotional regulation, motivation, and integrated functioning can result. The frontal cortex is the final region of the brain to develop, and it continues to form well into the third decade of life. It is this region of the brain that holds the prefrontal cortex, often referred to as the CEO of the brain, as it contains all the elements that enable and regulate executive function. This important area of the brain is responsible for attuned communication; decision making; the expression of personality; capabilities regarding the discerning of good/bad, better/best, and the like; differentiation between conflicting thoughts; goal-directed behavior; and what neuroscientists call "social control." Social control refers to the ability to manage urges that could lead to negative effects if not controlled. When these regions of the brain develop in healthy ways, they can work in concert to create an integrated, adaptive, and responsive system.

While we have more ways of studying the brain than ever, the nascent nature of the relationship between technology and brain health, coupled with the slow speed of our research cycle, makes it such that we simply can't keep up with how technology may be shaping the brain. There is emerging and existing research, but as our imaging technologies advance, existing findings are made obsolete. Nonetheless, a general knowledge of the physiological structure of the brain alongside careful analysis of the research about technology tells us there are, indeed, neurological impacts.

The very existence of neuromarketing as a field of study confirms this. Neuromarketing experts use brain-imaging technology along with biometric measures (heart rate, respiration) to determine why consumers make the decisions they do. By studying fMRI scans and other physiological data while individuals interact with technology, these researchers can see how activation of particular areas of the brain due to specific technological content exposure can result in specific behaviors, ideas, or feelings in subjects. By changing the way content is delivered within the digital framework, these researchers can, almost literally, change the way the brain is activated, thereby changing the lived experience of the subject. This whole endeavor is predicated on the knowledge that activation of certain brain regions will bring about certain responses. Given that the brain wires together where it fires together, these repetitive exposures and responses to technology must be having at least some impact on the way our brains are wiring.

Several decades ago we learned that lots of stimulation led to lots of complex wiring in the brain. This led to claims that television and other forms of high-stimulation media could *help* the brain. Programmers created "educational" shows and videos, but along with the educational media came a preponderance of overstimulating and not-created-for-educational-purposes programming. Ultimately, children began consuming an abundance of screen-based entertainment. Over time, high-quality research emerged suggesting that too much stimulation might be leading to an increase in ADHD in children. As a result, scientists began studying rats to see if screen exposure impacts the developing brain. The results were clear then and still are today: screen exposure shapes the brain, and the interpretations around these changes are far ranging, conflicting, and complex.

Here is a description of one such study: Neuroscientist Jan-Marino Ramirez, director of the Center for Integrative Brain Research at Seattle Children's Hospital, found that mice exposed to six hours of light and sound similar to a video game showed "dramatic changes everywhere in the brain." In a National Public Radio (NPR) interview, Ramirez noted: "Many of those changes suggest that you have a brain that is wired up at a much more baseline excited level. You need much more sensory stimulation to get [the brain's] attention." While he notes that the "wired" mice remained calmer in stressful situations, he also found that they manifested symptoms akin to ADHD, were prone to riskier behavior, and demonstrated learning problems.

Whereas Ramirez says these responses to video game–like exposure suggest prudence in exposing our own brains to the same, researcher Leah Krubitzer, an evolutionary neurobiologist at the University of California, Davis, interprets the findings differently. She asserts that these changes may be beneficial in that a less-sensitive brain may be adaptive in a world filled with overstimulation. Regardless of interpretation, however, most scientists agree that the brain can change in response to the demands of the environment. The preponderance of screens in our current environs is difficult to overlook, suggesting that they will impact neurological health. Ultimately, it is up to each of us to keep tabs on how we are adapting—and how these changes help or hurt us.

If you have been to one of my talks, you are familiar with how I demonstrate the way in which the brain adapts to its environment. First, I show the introduction and play the theme song from a 1969 episode of *Sesame Street*. The images are black and white, and each sustained camera shot lasts somewhere between six and fifteen seconds. If we know that the brain wires together where it fires together, we can assume that individuals who are exposed to this kind of pacing in the presentation of screen imagery will develop circuitry used to *waiting for up to fifteen seconds* for a new stimulus. Doing this over and over would force the brain to develop the ability to focus our attention without becoming bored or distracted for about that amount of time, fifteen seconds.

Next, I show the same introduction, but from a 1984 *Sesame Street* episode. Now the sustained camera shots last between three to six seconds, with a few lasting only one and a half seconds. The brain exposed to this kind of rapid cycling of stimulation and images doesn't wire with the same tendency toward focus and boredom tolerance that we explored earlier. Instead, it will anticipate a change of scenery *every three to five seconds*, wiring for efficiency in handling multiple images in fast succession.

Finally, I play the same *Sesame Street* introduction and song, but from the past few years. In this current rendition, there is no sustained, unmoving camera shot! Instead, images move constantly across the screen and scenes are built before our eyes. This means the brain is trained to expect *constant, changing stimulation.* If things don't change on the screen immediately, our brain is trained to look away to find something novel to attend to. When the preponderance of visual stimuli presented to us follows this pattern, over time, we no longer have the neurologically practiced skills of waiting and focus.

We may not only be changing the wiring of the brain in the regions related to focus and waiting by our exposure to constantly moving stimulation, but the games, media, and use patterns of such also impact the way in which our brain experiences reward. Dopamine is a neurotransmitter that helps control the brain's reward and pleasure centers. It helps us see the potential of reward as well as helps us move toward achievement. Dopamine is released with drug use, video game play, and many other situations that involve anticipation and reward. When dopamine levels are high, we feel a sense of pleasure. Once we've experienced the resulting feelings, it's hard to want to live with less.

Given the role and mechanics of dopamine, it's easy to understand that it is involved with most behaviors we may consider "addictive." The National Institute of Mental Health's Dr. Jay N. Giedd, chief of the Unit on Brain Imaging in Child Psychiatry, insightfully reports: "All of our basic drives (e.g., hunger, sex, sleep), all substances of abuse, and everything that may lead to addiction (i.e., compulsive behavior characterized by loss of control and continuation despite adverse consequences) increases dopamine in the nucleus accumbens." This means that our compulsive engagement with devices, especially those that are highly stimulating, leaves our brains bathed in dopamine and wanting more.

Interestingly, though, the clinical and research communities are conflicted about whether technology addiction is "formally diagnosable" in relation to internet use and gaming. To say the least, the point at which technology use becomes akin to addiction is a tricky one to discern. The American Psychological Association has chosen to exclude internet, gaming, and social media addiction from inclusion in their *Diagnostic and Statistical Manual*, and many researchers support this, saying they are hesitant to "pathologize" a behavior that is highly normalized and even necessary for daily life. There is emerging research, however, that supports the idea that internet use in general, and video gaming, social networking, and porn specifically, may have the potential of becoming addictive. Stanford researcher Elias Aboujaoude found that fMRI scans of individuals who are using internet-connected devices show activation of reward pathways in the brain similar to the fMRI scans of those who have substance abuse disorders. Tolerance, wherein people move from being content with a certain amount of use to needing more use to attain a previous level of stimulation or satisfaction, also supports an addiction model.

Regardless of whether internet addiction (gaming, shopping, social networking, or porn addiction) is recognized as a diagnosable affliction, we must take seriously the ways in which our bodies and minds can become dependent on devices and the digital worlds they provide. As tech companies and other entities that exist to serve us or sell us products become increasingly capable of recording our use and using the data to manipulate us and keep us "on screen," we must take charge of our own independence. It is crucial that we become conscious of and serious about the possibility of dependence or addiction in ourselves and our children. Some telltale signs include:

- Moving from incidental use to nearly constant use.
- Needing increasing levels of tech time or stimulation for satisfaction.
- Being jittery or anxious in response to stepping away from technology.
- Lying in order to garner more time/specific content/etc. or to cover up certain forms of use.
- Isolating in order to engage technology.

Whether these symptoms are related to psychological constructs such as depression or anxiety or to internet addiction or dependence may be difficult to know. Either way, however, it is important to consider how our technology use is helping or hindering our emotional well-being. We must also ask ourselves what our overdependence on screens is costing us in the way of developing our whole selves as relational, emotional, physical, and spiritual beings. If we are isolating in order to engage technology, we must pay special attention, as this doubly hurts us: we are not getting social practice, and we are also keeping ourselves from opportunities for it.

TAKE ACTION

Preventing Tech Addiction . . . and Getting Help

SET CLEAR BOUNDARIES, COMMUNICATE THEM, AND ENFORCE THEM.

This is self-explanatory and applies to oneself as well as to our children.

THINK AHEAD BEFORE ADDING A TECHNOLOGY.

Identify the particulars about the technology or platform that may lead you to slip into addictive-like patterns. Use this information to determine if it is wise for you to own the specific tech or app. If not, find other places to fulfill and address this interest of yours. If you choose to move forward with engaging the technology or app, set healthy and intentional norms around your engagement with it, and establish regular intervals to check how you're doing. Set alarms or note a check-in time on your calendar. For parents, before introducing new technology or apps to children, be sure to practice this same kind of consideration and help your children choose technologies and apps that don't fall into the addictive category.

MAKE SURE TECHNOLOGY IS NOT YOUR ONLY "SWEET SPOT."

We can become drug addicted as a way to cope with difficulty or to stimulate us when we don't know how to do so otherwise. The same can be true with technology. Making sure you have lots of stimulating and soothing things in your life that are not digitally driven will help you keep tech in its place.

INTRODUCE HIGH-QUALITY, SLOW-MOVING TECHNOLOGIES FIRST,
AND STICK WITH THEM AS LONG AS POSSIBLE.

Resist the habit of needing the newest, shiniest, fastest, and the best. Once you up the level of intensity or content in digital spaces, it's hard to go back. Our brains and bodies love the novelty and quick pace. With children, stay with slower-moving and higher-quality content as long as you can. When they want something newer, faster, and "shinier," have them write or draw a proposal for why it is healthy for them to have it. Make sure they address both potential positives and drawbacks. Have them articulate how they will keep their use in check.

IF YOU FEEL YOU'VE MOVED INTO USE PATTERNS THAT ARE HURTING YOU
OR KEEPING YOU FROM YOUR EMBODIED LIFE, GET HELP.

Many therapists can help with obsessive thoughts and compulsive behaviors. Even if internet addiction is not diagnosable, it is completely likely that obsessive thoughts about the ever-present digital world can lead to compulsive behaviors. When compulsive behaviors are approached as such, psychotherapies that help with obsessive compulsive patterns may be very helpful. Clinicians practicing acceptance and commitment therapy, cognitive behavioral therapy, and mindfulness meditation are great resources. If outpatient help is not enough, look for residential treatment programs. A call or email to ask about their accreditation and treatment approach will help you determine which is right for you.

TECH AND THE ISSUE OF MULTITASKING

In addition to being driven by the dopamine swimming in our synapses, we also are behaviorally influenced by the pace at which life moves. When we work at one task at a time, the frontal cortex is able to coordinate, and both halves of the brain work together to bring about goal completion. When a second, or third or fourth, task is introduced, the two halves of the frontal cortex begin to function independently, causing us to forget details and make mistakes. When we attempt to multitask, we aren't completing many different tasks with our full capabilities. Instead we are switching our attention from one task to another, often midway through each one.

Even still, "multitasking"—a fancy word for "distractibility" and frequently renamed "task switching"—is touted as a necessary skill. Studies by Stanford researchers show that individuals who claimed to multitask frequently and well were actually less effective than people who did one thing at a time. The researchers also found that subjects who were bombarded with several streams of electronic information were less able to recall information, pay attention, or complete tasks than the single taskers in the study. It is thought that the frequent multitaskers performed worse because they experienced difficulty organizing their thoughts and filtering out irrelevant information. They also were slower at switching from one task to another than those who completed one task before moving on to the next. About the self-proclaimed "successful multitaskers" who were found to be less effective, the lead researcher told a Stanford reporter, "They're suckers for irrelevancy. Everything distracts them."

Attempting to multitask doesn't only impact effectiveness, it also appears to lower IQ. Researchers at the University of London found that heavy multitaskers functioned from an IQ level approximately ten points lower than their actual IQ. One of the study's results suggests that for the average adult male, multitasking while trying to compose an email made it such that he might as well have let an eight-year-old write the memo for him. Furthermore, researchers at the University of Sussex found that participants who reported using several devices concurrently had less gray matter in the anterior cingulate cortex (the part of the brain responsible for cognitive and emotional control functions, including empathy) than subjects who used only one device at a time. While this correlation cannot infer causation, it is an important finding for directing further research and our thinking about the brain and technology.

It isn't just multitasking that reflects behavioral choices that may reflect changes in the brain. Researchers assessing the MRI scans of individuals who reported playing first-person shooter games such as *Call of Duty*, *Killzone,* and *Borderlands* 2 for nineteen hours a week found they had less gray matter in the hippocampus than the non-gamers in the study. The hippocampus is important for spatial navigation, stress regulation, and memory. To see if they could re-create this finding, the researchers then asked non-gamers to play either a first-person shooter game or a *Super Mario Brothers* game in which a plumber tries to rescue a princess. The results were clear: the action gamers showed reduction in the gray matter in the hippocampus; the *Super Mario Brothers* players gained it. It may be that this is because action games overlay visual geographic and mapping cues that let the brain ignore the need to pay attention to and critically consider one's surroundings in the same way that less action-oriented games do. The study's lead researcher, Gregory West of the University of Montreal, summed up the findings by saying, "While we train up this one system, this other system is potentially being neglected and potentially showing signs of atrophy."

THE GOOD NEWS OF NEUROPLASTICITY AND DOING "DEEP WORK"

These findings all point to the fact that there is a "use it or lose it" principle at work in our brain development and function. When we don't expose ourselves to opportunities for sustained focus, we literally prune off the neurological wiring that would allow it. When we train ourselves to task switch or expose ourselves to constant, quick cognitive input, we don't develop the circuitry to do the deeper work of navigation, focused thought, and complex analysis. Put a young child who is used to constantly changing visual, auditory, and neurological stimulation in a classroom with one teacher who speaks for longer than three minutes, and that child is literally at a loss for how to maintain focus. The same is true for most of us.

Thankfully, one of the most amazing realities of our humanity is the fact that our brains can heal. Referred to as "neuroplasticity," the brain can actually redirect wiring. With deliberate training and practice, we can reactivate regions previously pruned off. This means we can work to develop new wiring in the frontal cortex of the brain by exposing ourselves to situations that require waiting. We can create a greater capacity for emotional regulation by putting ourselves in situations that stimulate our feelings and then working to find ways to understand, express, and regulate them. We can learn to focus our attention and maintain deliberate awareness by creating strategies and practicing. Exposure and practice like this can, quite literally, create the kind of complexity in our brains that leads to a greater level of emotional and cognitive regulation in our lives. This, in turn, makes us healthier and more-grounded people. (Refer to chapter 10 for more information about how technology disrupts the development of focus, delay, and self-regulation—and how to cultivate these skills in our online age.)

In his book *Deep Work: Rules for Focused Success in a Distracted World*, Cal Newport writes about a social media pioneer buying a round-trip business-class ticket between the United States and Japan so that he could be free of distractions and write his book. Since that time, even this "place of peace" no longer exists, as planes, as well as nearly all other forms of transportation, offer unlimited internet access. Newport's wisdom, however, is more relevant than ever, as he argues that the disappearing skill of focusing on a single task without distraction allows for both higher levels of mastery and quality of outcome. "Deep work," as he describes it, involves intense concentration for significant periods of time in the absence of distractions. Focus, for Newport, is the new IQ in the information economy, and its mastery results not only in a higher level of craftspersonship but also gives us a greater sense of accomplishment.

I believe most of us wish we could do deep work. I have yet to meet someone who enjoys the ambient anxiety that comes from consciously stepping away from all the alerts, information, opportunities, and experiences that our devices deliver. The difficulty is that we now live in a space where our behavioral realities and the structure of our brains preclude us from the skills necessary for tuning out distractions. The good news is that we can change this.

An incredibly potent tool for learning to deal with distractions is mindfulness meditation. When practiced regularly, this form of meditation effectively helps us learn how to create a pause between an impulse and response. Basically, it makes us capable of being present to any given moment in a regulated state. It also works at a neurological level. Harvard-affiliated researchers at Massachusetts General Hospital found that an eight-week mindfulness meditation practice of thirty minutes a day altered the gray matter density in regions of the brain related to memory, sense of self, empathy, and stress responsiveness. Researchers at UCLA found that gray matter in the limbic and frontal cortex of longtime meditators was significantly more dense than in non-meditators. More recently, other studies indicate that long-term meditators have increased gyrification, or "folding," of the cortex, which may allow the brain to process information faster.

Taken together, this means that individuals who are practiced at being still, noticing their current state of being and directing their attention with intention, succeed in wiring their brains for functions such as emotional regulation, focus, attentiveness, and self-awareness. All these skills are at risk when we are distracted and overstimulated by our devices. When we practice the ability to be with our own selves in focused and still ways, however, we are basically redirecting brain activity from the more ancient, reactionary portions of our brain (the brain stem and some portions of the limbic system) to the more rational centers of the brain in the prefrontal cortex. In so doing, we train ourselves to fall less often into fight, flight, freeze, or faint mode and to rely more on the executive function skills to which the frontal cortex offers access. While there are forms of digital stimulation that can lead to an increase of gray matter in the frontal cortex (certain video games, for instance), the ability of meditation to impact gray matter while, at the same time, "unlinking" the brain from its reactionary stress response (located in the brain stem and limbic system) makes it powerfully unique.

Even ten minutes a day can help us create this powerful access. I'll go into this at length in chapter 11, but for now, some basic ways to take charge of our brain health would be to force ourselves, at least some of the time, to turn off all notifications, to walk away from our devices, and to force ourselves to focus by practicing mindfulness meditation. Creating a comfort with boredom and cultivating creative responses to it also does much to foster a sense of openness to the present moment, which enables us to tolerate the possible consequences of taking intellectual, emotional, and relational risks. When we know we can come back to ourselves in a calm way, pausing to care for ourselves when dysregulated, we can risk trying new things. When we know we can tolerate boredom without becoming anxious, we can create spaces free of overstimulation. These spaces, in turn, offer us an ability to focus and do deep work, which will improve both our outcomes and our sense of accomplishment. By exposing our brains to slower-moving stimuli (and, some of the time, to nothing at all), we allow it to wire itself in a way that facilitates deeper thinking and complexity.

VIOLENCE AND PORNOGRAPHY . . . AND THE BRAIN AND BODY

We cannot discuss the impact of technology on our brains and bodies without addressing another important topic: exposure to violent and sexually explicit media content and how it may impact our brains and bodies. While research is constantly evolving in this area, there are a few important themes to note.

For decades, researchers have been trying to understand the relationship between exposure to violence and aggression in media. Researchers in Denmark published an editorial-style review of what we know about this topic to date, and Lisa Borsellino expertly summarized their findings for the *Medical News Bulletin.* In a nutshell, although there are a few outliers, most of the data collected throughout the world has been fairly consistent: cross-sectional studies have established that violent media consumption is associated with pro-violence attitudes and aggressive behavior. In fact, studies are no longer looking at *whether* violence in the media increases aggression but *how* it does so. Similarly, experimental studies—those that can demonstrate whether the condition actually causes the variable—have shown a causal link between violence in media and aggression, and longitudinal studies are pointing toward a mechanism for the increased aggression. Given the reality of ever-changing forms of violent media content and the systems that deliver them, it may be impossible to know exactly how exposure impacts us physiologically, neurologically, and behaviorally. Current research, however, suggests that there certainly are effects—and we would be wise to take this into consideration as we shape the patterns of our digital engagement.

It would be irresponsible to talk about how we are shaped by technology and not discuss porn. Pornhub analytics tell us that 92 billion videos were watched in 2016 alone. This adds up to 4,599,000,000 hours spent watching porn. This is 5,246 centuries worth of time and is up 300 million hours from the year before. In 2016, Pornhub recorded 64 million visitors each day. That computes to 729 visitors every second. Average users spend nine minutes and thirty-eight seconds on the site, and this number has increased every year in the past five years. The United States scores highest in per capita consumption internationally, and millennials make up 60 percent of viewership internationally. Use by women is at 26 percent. Numbers like this cannot lie. As humans, we are highly invested in pornography use. As virtual reality (VR) explodes onto the scene, I believe this will become an even greater reality. When Pornhub announced its VR channel in 2016, it gave away ten thousand free VR glasses in twenty-four hours; by the end of the year, viewership went up 302 percent!

There are few subjects rifer with competing research findings and speculation than that of internet porn. Given porn's rapid and pervasive rise in accessibility and delivery systems, the massive amounts of money put into its development and promotion, and the anonymity users have in accessing it, more and more porn is being accessed by younger and younger individuals. While research surrounding possible impacts of exposure to porn among adults is emerging, research regarding more vulnerable populations, such as children, is fraught with ethical and practical difficulties. Given that the brain develops crucial parts of the frontal cortex that are responsible for relational attunement through early adulthood, the fact that high amounts of graphic pornography are being consumed by individuals in late childhood and early adolescence, and that this content is highly provocative to the brain, is important to consider.

While some claim that greater access to pornography has a positive benefit of enhancing sexual education, research has found that porn exposure in adolescence may contribute to the development of unrealistic values and beliefs and result in earlier and more promiscuous sexual behavior patterns and sexual experimentation, as well as sexual preoccupation. In addition, a link between self-concept and porn use has emerged in literature, with girls feeling as though they can't measure up to the images of women they see in porn and with boys experiencing concern about not being able to perform in keeping with forms of sex delivered by most porn platforms. The researchers end their report by noting that "adolescents who use [internet] pornography . . . have lower degrees of social integration, increases in conduct problems, higher levels of delinquent behavior, higher incidence of depressive symptoms, and decreased emotional bonding with caregivers." More research needs to be conducted to determine if these findings hold.

One finding that has been shown to be reliable is that violence and sex are highly relevant to the human brain's reward system, especially for the adolescent brain that is constantly looking for novel and provocative stimuli. Given that much of today's porn explores the pain-pleasure binary and includes violence as a medium for this, it is important to discuss this with youth before they become comfortable with porn based on infliction of pain. It is imperative that we establish trusting relationships with ourselves and with one another so that we can talk honestly, candidly, and in non-shaming ways about violence and highly provocative, sometimes violent, sexual content. We do not benefit by putting our heads in the sand or by making ill-informed, overly simplistic ultimatums.

An important and relevant dynamic of current porn use is the manner in which it is consumed. As stated previously, users in 2016 spent an average of nine minutes and thirty-six seconds on the Pornhub site. In that time they cycled through anywhere from eight to fourteen different sites. This means that, just as our earlier description of never-static images makes an impact on our ability to tolerate one image at a time, we are linking constant new stimulation to the ideas and behaviors around sex. In a 2012 interview with *Playboy* magazine, singer-songwriter John Mayer reported,

"Pornography? It's a new synaptic pathway. You wake up in the morning, open a thumbnail page, and it leads to a Pandora's box of visuals. There have probably been days when I saw 300 vaginas before I got out of bed. How could you be constantly synthesizing an orgasm based on dozens of shots? You're looking for the one photo out of 100 you swear is going to be the one you finish to, and you still don't finish. . . . How does that not affect the psychology of having a relationship with somebody? It's got to." While Mayer may not be a neuroscientist, he brings home an important truth: the way in which we view porn impacts our embodied selves and relationships. Stay tuned, because researchers are beginning to study the sharp rise of sexual dysfunction and low libido in men under the age of forty (most remarkably, men under the age of twenty-five make up a large portion of this group) who consume internet porn.

Sadly, I believe it is impossible to keep ourselves and our children from encountering violent and sexual content. It used to be common thought that exposure to media violence was not correlated with criminal violence. It also used to be thought that exposure to porn could enhance one's sexual function. We used to talk about setting up relatively simple filtering on our computers to protect our children from both. None of these things seem relevant or straightforward anymore. Even soft forms of violence, when repeatedly engaged digitally, can accumulate to create a comfort with aggression, and even limited contact with porn could cause sexual side effects for some.

There is no one-size-fits-all approach to dealing with violence and porn. We must become discerning consumers and must work to stay informed about new platforms and places and how they might impact us. Furthermore, we must support well-constructed research to understand how violence and overly sexualized content impacts the development of children and assist in creating a world where this can be minimized. At the very least, we can be careful about our own support of platforms that glamorize violence and twist the beauty that is sex into something less than what it is meant to be. And we can initiate non-shaming conversations about both.

AN IMPORTANT CONSIDERATION REGARDING RADIATION

One final topic pertinent to this chapter is radio-frequency radiation. At the time this book is going to press, new data is coming out about exposure to radiation emitted from the electromagnetic fields (EMFs) of digital devices. In May 2016, the US National Toxicology Program, a program of the National Institutes of Health, released the results of a two-year study that found that rats exposed to high levels of radio-frequency radiation developed cancerous tumors. Research with humans, however, has been mixed. Some research shows no effect at all, while some shows a heightened risk of a specific type of brain tumor with excess exposure to the nonionizing radiation emitted from cell phones.

While analysis of the data on the rats studied in the NIH experiment is not yet complete, the American Academy of Pediatrics feels it provides important reminders to limit the interaction children have with digital devices and to treat cell phones as tools, not toys. Children are still developing and so are at particular risk from early exposure to the radiation emitted from digital devices. This must be tended to as the research emerges. It is always better to be safe than sorry!

TAKE ACTION

The American Academy of Pediatrics makes the following suggestions for reducing the risk of radiation while using devices:

- Text message more often than voice calling on a cell phone; when using cell phones, do so in speaker mode or with the use of hands-free kits.
- When talking on the cell phone, try holding it an inch or more away from your head.
- Make only short or essential calls on cell phones.
- Avoid carrying your phone against the body (as in a pocket, sock, or bra). Cell phone manufacturers can't guarantee that the amount of radiation you're absorbing will be at a safe level.
- If you plan to watch a movie on your device, download it first, then switch to airplane mode while you watch in order to avoid unnecessary radiation exposure.
- Keep an eye on your signal strength (i.e., how many bars you have). The weaker your cell signal, the harder your phone has to work and the more radiation it gives off. It's better to wait until you have a stronger signal before using your device.
- Avoid making calls in cars, elevators, trains, and buses. The cell phone works harder to get a signal through metal, so the power level increases.
- Remember that cell phones are not toys or teething items.

5

Technology and Relationships

Almost every person I meet has some sort of horror or glory story about how technology has impacted his or her relationships. This is the one topic on which most people agree: technology *has* changed both the who and how of our relational experiences. Not only has technology given us access to an infinitely larger pool of people with whom to connect, but it also has provided an endless array of options about the way in which we establish and maintain those connections. People now routinely encounter new friends and potential romantic partners in a plethora of constantly changing digital meeting spaces. Gamers join clans of players from around the globe and invest deeply in shared strategy implementation and play. Students take classes from professors on other continents. Teams of coworkers congregate in digital conference rooms and share their screens from the other side of the world. Support groups and therapy relationships exist via email and texting threads. We all have more relational potential than we'e ever had!

The benefits of the expansion of our relational options are real. Never before have we experienced such unbounded opportunities to have meaningful personal encounters with people regardless of geographic location or other variables. Once connections are made, we can maintain and even deepen relationships with a host of communication and connection tools heretofore unavailable. These and many other positive outcomes of technology cannot be denied. Even still, the way in which technology is shaping the scaffolding upon which we build our relational lives is complex and worth taking time to consider.

THE RELATIONSHIP WITH ONE'S SELF

When speaking of human relationships, it's easy to fall into the trap of starting with a discussion of our relationships as individuals with other individuals. While this makes sense, no discussion of relational health is complete without acknowledging the structural significance of our own internal relationship of self to self. Much of what I mean by this will be covered in the next chapter. For now, however, before launching into a discussion of how technology impacts the nature and quality of our relationships with others, it's important to acknowledge that all human relationships are impacted by the sturdiness of the individuals involved. If I am not fully in touch or in harmony with most of the disparate parts that make up "me," it's unlikely that I'll be able to engage in a fully honest and healthy relationship with you.

This is important because at the same time technology is impacting how we relate to one another, it also is having a profound effect on the way we relate to our very selves. If I have been able to develop empathy and a stable internal locus of control alongside my tech engagement, then I will be able to build an authentic, deep relationship with you. If, however, the prevailing nature of technology's impact on my relationship with myself has been to make me self-promoting, self-centric, lacking in empathy, limited in my communication skills, and without an accompanying sense of self-knowing awareness of my limitations as well as my strengths, then my relationship with you will be built on a fragile foundation. Keeping this foundational dynamic in mind is critical as we discuss our relationships with others.

RELATIONSHIPS WITH OUR DEVICES

A second foundational reality is also important to note: given the amount of time we spend with screens, it seems plausible to posit that some of our most meaningful relationships now exist with our devices (if meaningfulness is, at least in part, determined by investment of time and energy). While these devices serve to connect us with people, we are, over time, developing response patterns to devices that look much like our response patterns to humans. Think about how we feel "lost" when we forget our phones and panic at the thought of going without them. Research has shown, in fact, that interaction with our devices can stimulate the release of oxytocin, thus initiating feelings similar to love. Considered the "cuddle hormone," oxytocin is similarly released when a new mother lovingly gazes at her nursing baby. In short, our physiological responses to devices suggest an emotional connection to them not unlike what we experience as physiological responses to connection between humans.

The preponderance of digital "personal assistants" is a potent example of this. These devices are capable of voice interaction, music playback, making to-do lists, setting alarms, streaming podcasts, and providing weather, traffic, and other real-time information as well as playing games such as *Jeopardy!* The epitome of these personal assistants can be found in the blockbuster hit *Her*. The film follows a man (Joaquin Phoenix) who develops a deep and complicated relationship with Samantha, an intelligent computer-operating system personified through a female voice (Scarlett Johansson).

In truth, we are all vulnerable to this kind of emotional connection to objects. As augmented reality and artificial intelligence (AI) offerings become more sophisticated and mainstream, we are finding ourselves with ever-increasing opportunities to interact with devices as "friends" and others to whom we relate. When we do this with repeated consistency, it doesn't seem far-fetched to imagine the emergence of actual and complex feelings toward our devices, even when we know they are not "real." Similarly, it follows that our human relationships will subsequently be affected. Our human partners and friends simply cannot be as ever present, as effectively able to predict our wants and needs, or as "low need" as our digital devices can be.

IMPORTANT DISTINCTIONS BETWEEN RELATIONAL LIVES AND SOCIAL LIVES

It is my firm belief that humans are, primarily, relational animals. We are built for and by connection, and the nature and the quality of those connections contribute to all manner of personal qualities, skills, and potentials. Our early relationships in particular—those with our parents or caregivers in the first six or seven years of life—influence our most basic interpersonal relational patterns and propensities throughout our lives. When these early connections are reliable, loving, and appropriately oriented, we develop into people who can bring our whole selves into relationships with others without becoming enmeshed or rejecting. We can tolerate moving from states of intimate connection to independence, we can attach and we can separate, and we can form interdependent relationships that are fluid and diverse—while, at the same time, maintaining a solid and flexible sense of self. We don't need people to conform to our standards or preferences to relate to them, and we can maintain the integrity of ourselves even in conflict or disagreement.

The continued development of this kind of healthy relational life is dependent on intentional work, commitment, and practice. It involves the active pursuit of meaningful communication with self and others as well as effort exerted toward behaviors intended to build and maintain relationships. In other words, practice, practice, practice. This is well illuminated by Aziz Ansari and Eric Klinenberg in their book, *Modern Romance*. The authors speak of reimagining online dating sites as online "introduction services." They wisely note that when interactions executed within such platforms are labeled "dating," we place far more weight on them than would be wise. When, however, we consider them places merely for being introduced to people we may or may not choose to spend time with in embodied spaces, we keep them in their rightful place. In other words, if we've responded to the winks and right swipes of potential suitors but not dated them, we've simply *interacted* with them. If we consider every botched or less-than-dynamic interaction as a tally mark, we are at risk of telling ourselves that we've completely failed at dating, that we are unlovable, and that there's no one out there for us. It's a completely different matter altogether if we only count those interactions wherein we've intentionally spent time together in embodied space. One is about screening, gut reactions, and clicks on a keyboard. The other is about interpersonal engagement and process.

We do a less-than-stellar job of making these kinds of distinctions all the time. We say we've "talked to" someone when we've merely texted with that person. We consider our online contacts "colleagues," even if we have no idea who they are or how we originally became connected. Perhaps the place we blur the lines most is in our socially networked spaces where we consider ourselves "well-friended" or not.

To truly be honest and authentic, it would behoove us to make a distinction between our social lives and our relational lives. The former refers to the relative *amount of time* we spend in companionable connection with others, whereas the latter refers to the *part of our beings we invest* in knowing others and being known by them. To be healthy, lasting, and reliable, these relational forms of knowing are best predicated on communication that is honest and authentic, happens within a variety of contexts, and occurs over time. This relational communication ultimately moves individuals through an intentional process of disclosure, listening, and discernment of trust that constitutes the developmental process of progressing from acquaintance to friend.

For the purposes of our discussion, consider what, if any, difference exists between the relational lives versus the social lives of an individual. In my mind, intentional embodied relationships exist along a continuum from casual to intimate. Although social relationships exist along a similar continuum, an inherent difference exists between engaging in a mutually sought after and maintained relationship with a person and being exclusively "social" with that person. We might socially interact with many in a shared space but only have actual *intentional* relationships with a few.

As I interact with all kinds of people at talks, in my office, and as I move through the world, I see firsthand that the lines between our social lives and our relational lives, and the health and depth of each, are more blurred than ever. I repeatedly encounter individuals who have social network profiles packed with people, yet they still feel lonely and alone. In fact, research shows that the use of multiple social networks increases the likelihood of depression and anxiety and that even a single negative experience on Facebook can put a person at risk of depression (according to 80 percent of respondents). Psychologist Jean Twenge, in her generational research, finds that social media use is linked with higher levels of depression and suicide and lower levels of happiness in teens. In her view, the twin rise of the smartphone and social media has created a reality of unhappiness previously unseen in adolescent populations. While it is not possible to know whether broad social network use causes depression or if people who suffer from depression are more prone to wide social network engagement, the correlation exists.

It isn't just our emotional reaction to social and relational networks that is important. What requires equal reflection are the assumptions we make about the fullness or emptiness of our relational quiver based on "friends," "likes," and "followers." While the number of friends one amasses in digital spaces presumably corresponds with real people, it is unclear if these connections realistically and reliably have what is required to maintain the kind of embodied support we all need to get through life. While the automatic assumption may be that we are "sharing life" by posting in social networks, the veracity of online connections may be very strong in some cases and virtually nonexistent in others, leaving us unclear as to what information is actually tended to by one's social networks. Furthermore, the depth of support can be tricky to ascertain in such places. A simple "heart" response or "thumbs up" may mean much to the responder but leave the person posting still alone in facing the reality about which he or she has posted.

Basically, it's unclear whether people with bodies and physically manifested lives can be entirely relationally fulfilled by connections made and maintained solely in digital spaces. Given our mass move toward connection moderated within these spaces, however, it would seem as though we believe that we can. If we were to explore the possibility that online social networks don't exclusively contribute to heartier and healthier embodied relational networks, what might we find? How might we consider why we engage them with such fierceness and regularity?

Perhaps we might find that building and engaging social networks helps us feel as though we are doing something to feather our embodied relational nests. I often use two analogies to describe the way in which we feel that tending to our social selves will address our relational needs. The first of these begins with me at a starting line, beginning a running race ill-informed or prepared for what I will need to complete the race with health. If I haven't eaten, my stomach is growling and empty, and the race is about to begin, I can quickly grab my water bottle and chug its contents. My stomach will feel full and stop growling, but somewhere, very early on, my body will realize it doesn't have the calories it needs to support the exertion required to finish the race. The second analogy has to do with the nets under high-wire acts in modern-day circuses. If I were on a high wire, I would want to be able to look down and see a huge, wide net prepared to catch me in case of a fall. Modern-day circus performers, however, see narrow nets, barely visible to the audience. While they are not much wider than the body width of the performers, those nets are strategically designed and placed and have the structural integrity to support the necessary weight and velocity of a falling performer.

Each of these examples points out that simply filling in space is very different from determining what is specifically needed in a space and how that need might be anticipated and met. In our high-demand world, it is likely that we will need very specific kinds of "nutrients" and "supports" to function with health and excellence. When it comes to embodied relationships, we need intentionally created and regularly maintained diversity in our relational circles. We need at least some relationships of emotional depth among our many social connections. If we spend all our time and energy building social networks and very little of it creating and maintaining embodied connections, we may not have what we need when we run the race or make a misstep and fall. In our relationships in digital spaces, as in our relationships in the embodied world, the strength and resources available to us when we need them are directly related to our investment and authenticity in developing them. Until we live all our lives in digital spaces, we need at least some embodied people upon whom we can rely and who can, in turn, rely on us. Given this truth, at least some of our time needs to be spent creating and investing in relationships with embodied others. No matter how uncomfortable this is, we must make these relationships a priority.

TAKE ACTION

Consider what you believe to be the differences and similarities between social networks and relational connections.
Try these approaches:

- Track the number of responses to social media posts you make in a day and compare that to the number of texts or phone calls you make.
- Consider who you might call if you had an amazing piece of news to share or if you needed help in an emergency.
- Let the difference between the types of connections and relationships you enjoy sink in, and determine where you might make some investments to deepen those that have real potential.

STATUS, COMPARISON, AND FOMO (FEAR OF MISSING OUT)

A few of the hallmarks of the nature of the relationship (or lack thereof) between our relational lives and social lives are our comfort with authenticity, our relationship with the concept of status, and our fear of missing out. For individuals who are less prone to compare their own circumstances or selves with those of others or who feel relatively satisfied with their relationships in general, engagement with social media may have few to no adverse effects. If, however, we frequently feel "less than," find ourselves comparing ourselves to others, or feel discontent with the nature or quality of our life or relationships, social networks can be a minefield.

My engagement with the research and thousands of people has led me to believe that social networks create a sort of constant background noise for those who engage them. Fed never-ending streams of status updates, we quickly lose time simply scanning through the curated presentations of our friends' lives. Gone are the days where we met people (in writing, by voice on the phone, or in embodied space), received a bit of information about them, shared some information about ourself, listened to our gut about what we learned (perhaps their level of interest in us and their trustworthiness), and then proceeded to gain and give more over time. Now we live in a world of fully curated and carefully edited disclosure at the click of a button. Frequently, we have amassed data about people long before we've met them in any kind of mutual exchange. This predicates our meeting upon what we have inferred based on what has been intentionally shared with a wide audience, which often cuts us off from the dynamic kind of paced uncovering that can come with embodied relationships deepened by discoveries made over time.

If all our sharing in social media spaces were truly honest, or if our profiles presented reliably and reasonably authentic and well-rounded images of ourselves across the board, this might be fine. Given, however, that this collected data is not about trying to present who we are in well-rounded ways but is rather a conglomeration of information we have chosen to share as "status," it frequently portrays only a portion of who we are or, in some cases, who we have created ourselves to be. Whether we have done so consciously or unconsciously, we are faced with a difficult reality: we must navigate the space between our social selves and our relational selves in this scenario, having amassed a following of digital friends based on a self-selected presentation of data that may not be found accurate when people encounter us in embodied spaces. This can create serious anxiety, even fear of rejection for who we really are.

Another reality of access to a twenty-four-hour-a-day feed of the status updates of others is the sense that we are forever missing out. This is something I hear consistently from individuals who interact with social media—and it almost always correlates to anxiety about missing both information and events, as well as about missing out on personal qualities that might make one more desirable. As I listen to people describe these phenomena, I hear themes similar to other situations in which fear is incited. We become hypervigilant, noticing every way in which we are ill-informed, inadequate, or overlooked. We may become reactive and strive to remedy our lack, succumb to feelings of powerlessness, or both!

It would be easy to feel as though I am criticizing those who engage social media. I am not. My intent is to incite and promote a sense of self-awareness about how much we rely on social media to provide substance in our relational lives (unless we are working actively to use it in that way specifically). I can relate to people who live with a curated online "self" about whom they feel ambivalent. Because our digital fingerprints and footpaths live well outside our control, it is easy to feel as though they have gotten away from us, even if we have worked hard to engage them well and keep them authentic. For individuals who have made missteps and have taken responsible steps to remedy them, a constant and very real sense of fear and dread exists when they know that people are checking them out online.

Similar levels of anxiety exist for anyone who has ever been bullied or had images or words posted about him or her that are less than flattering. Once something has been posted and a person has been tagged, it's nearly impossible to make it go away. For those who have lived through their tumultuous preteen and adolescent years online, there is a particular anticipation of being "found out" as they move into adulthood. If first impressions are important and the initial point of introduction has moved online, we all live largely out of control of people's first (and possibly lasting) ideas about us.

Even for individuals who strive to have a cohesive and authentic presentation of self in both embodied and digital spaces and who have been privileged to have nothing negative or untrue posted about them, engaging with online social platforms can be challenging. Many of these people maintain two or three online personas with entirely separate social media presences representing each. Others instead choose to represent their full self in one integrated space, often leaving many people from the divergent parts of their lives confused, even judgmental, about what they find in a feed but do not recognize or understand. It is a tricky relational world out there, and it takes a whole lot of time and energy—as well as a wide-eyed recognition of how little one actually controls the thoughts and feelings of others—to maintain relationships in the digital world. This work is not for the weak of heart, nor is it easy for the developing self. Those with developing hearts and identities (all teens and young adults) are faced with some of the greatest difficulties with online relationships. Perhaps this is because we are changing the very way in which we communicate without an ability to know the consequences of doing so.

COMMUNICATION TOOLS AS BUILDING BLOCKS OF RELATIONSHIPS

If our relationship with our own self and the authenticity of communication regarding that self is the foundation upon which our relationships are built, then the nature and quality of our *communication* creates the building blocks of our relationships with others. Research conducted with pairs of close friends found that communication via instant messaging results in significantly lower levels of bonding than face-to-face communication, video chatting, and audio chatting. If this is true for existing close friends, how might it impact the many relationships begun and maintained solely through typed digital messages?

Given the sheer amount of time we now spend with screens, it would make sense that we invest less time in practicing face-to-face and embodied person-to-person communication than we have in the past. Even while communication arts degrees grow in popularity, the tendency of the general public to invest in honing the effectiveness of their communication skills (written, verbal, in person, and online) seems to be on the decline. For this reason, we need to consider how our patterns of being in or out of touch with those in our lives might color our experience of relationships. Just as chugging a bottle of water can quell a growling empty stomach, making it feel full, so can shooting off a series of texts serve to fill in relational gaps. If we are seeking simply to send information, impart a sentiment, or collect responses this might be fine. If, however, we are reaching out for a deeper response, it may desperately miss the mark. Similarly, if we hope to maintain the kind of intimate and lasting connections that give us the necessary support to make it through life, we need to make sure we are practicing and maintaining an ability to communicate in deeper and more intimate manners. To do so, we must tell ourselves the truth about how our interactions in digital spaces may shape our unconscious assumptions about communication in general and how they might lead us to act in ways that don't get us what we need or want.

One of the potential issues with the digital world diminishing our communication skills is "disinhibition." As we spend less and less time practicing the art of communication, with its subtleties of give and take, we are noticeably shifting toward disinhibition, a lack of restraint that manifests in impulsivity, poor risk assessment, and a disregard for social conventions. This shift is particularly apparent in our typed communiques. In his article about "online disinhibition effect," Rider University communications professor John Suler describes how digital communication can train us to be less "other aware." He writes: "In text communication such as e-mail, chat, blogs, and instant messaging, others may know a great deal about who you are. However, they still can't see or hear you—and you can't see or hear them. Even with everyone's identity visible, the opportunity to be physically invisible amplifies the 'disinhibition effect.' You don't have to worry about how *you* look or sound when you say (type) something. You don't have to worry about how *others* look or sound when you say something. Seeing a frown, a shaking head, a sigh, a bored expression, and many other subtle and not-so-subtle signs of disapproval or indifference can slam the breaks on what people are willing to express. In [Freudian-based] psychoanalysis, the analyst sits behind the patient to remain a physically ambiguous figure, without revealing any body language or facial expression, so that the patient has free range to discuss whatever he or she wants, without feeling inhibited by how the analyst is physically reacting. In everyday relationships, people sometimes avert their eyes when discussing something personal and emotional. It's easier not to look into the other's face. Text communication offers a built-in opportunity to keep one's eyes averted."

This is important if relationships are built by two or more individuals disclosing and receiving information and feedback (verbal, nonverbal, contextual, and specific) in the course of an interchange. If we become that which we repeat, all our communicating outside the context of embodied others may be making us more self-directed and self-aware in our communication patterns. This can lead us to become unfamiliar or uncomfortable with the inherent vulnerability of person-to-person encounters, thereby avoiding embodied people in deference to increased communication in more anonymous or "eyes-averted" ways.

Interesting research evaluated whether subjects preferred to answer questions posed by humans or by "embodied conversational agents" (ECAs), which are virtual people. The results revealed that the study participants preferred speaking with ECAs if the answers might be of a highly sensitive nature or likely to involve negative self-admissions. If the answers were considered less sensitive or more likely to include positive self-admissions, the participants preferred human interviewers. The reason for this? Research subjects reportedly appreciated the lack of judgment an ECA would afford.

We practice this finding in ways large and small every day. When we don't want to have to deal with judgment, don't want to take the time to listen in return, or are insecure about our communication skills in general, we choose the least embodied form of communication. If we come to live by the behaviors we practice, it seems safe to assume that our texting, emailing, posting, and consuming of posts—all divorced from the presence of an embodied self or other—may be making us more self-centric communicators who are out of practice in the give and take of embodied verbal exchange. Given that communication forms and sustains relationships, it is wise to explore and examine this personally.

When we spend less time navigating the awkward moments inherent in embodied face-to-face interaction, we miss out on cultivating our communication skills. Without these skills, we feel inadequately prepared to handle the unknowns of person-to-person encounters. A cycle ensues, and our ability to initiate and maintain actual encounter and interaction suffers. When we miss even low-stakes opportunities for conversational practice (such as when we are standing in line) because we engage with our devices rather than with those with whom we are standing, we bypass opportunities that might help prepare us for situations when we really need to be able to interact with those around us. It is imperative we believe this! Truly, the more we practice communication in person, the better our communication skills will become. Being able to tolerate eye contact, to maintain discourse, and to have conversations about both nothing in particular and things very important benefits us in life in tremendous ways. Keeping these skills in tune may be more critical and require more intentional work than ever.

OBJECTIFICATION AND RELATIONAL SHOPPING

As our communication occurs increasingly in digital formats and in net-worked spaces, one of the subtler shifts is a tendency toward objectifica-tion. When we interact with curated representations of people (profiles) more often than with the actual embodied humans themselves, it is easy to develop a sense of distancing. That is, when we relate to others, even if only partially, as potential "likers" of our status or as status-building blocks for ourselves (as when we manage to acquire a "friend" who will gain us status with our online community), we objectify those people in subtle ways. From here it is a short step toward "shopping" for what we desire relationally from a never-ending stream of people and objects we know based on presentation of data only.

This ability to sidestep the messiness of encountering and learning to relate to people in all their authentic complexity can lead us to become under-skilled in actual relationship building. When we choose to connect to another based solely on the parts of ourself that we disclose via digital domains, we can choose, in many ways, to avoid any data *not* included. When we build an entire social network like this, we forgo the ability to learn to navigate differences, welcome diversity, and work through the complexities inherent in authentic connection.

When creating an online network, if we feel we are missing important pieces to a full relational pie, we can more easily "shop" for what we desire more than we could ever do in embodied spaces. Treating people as objects represented by their own social media presences, we can appease our sense of desire for adventure or diversity simply by sending a friend request. This does not grow our actual ability to engage others who are different from us and may, in fact, keep us hostage to avoiding relational risks in our embodied lives.

Nowhere is the tendency toward objectification and shopping more real than in online dating sites that rely on quick, split-second decisions regarding whether or not a person is interested in another person based upon a photo and tagline alone. Reducing people to an image (or two), or even to the qualities and interests they list online, disrespects the inherent complexity and value of every person. We seriously minimize the vibrancy of human relationship and the dynamic qualities that draw people toward each other in most online dating platforms and other social networks like them. To encourage healthy relationships, this must change.

The need to stand out alongside the constant comparison and competi-tion for attention in our socially networked spaces has the power to subtly impact the way we think about ourselves and others. In talking with many people who have completed media or technology fasts, I consistently hear that somewhere around day two of the fast, they become acutely aware of a judgmentalism and harshness in their internal dialogue, both about themselves and the world. In stepping back from their engagement with technology, they notice how loud and constant the commentary regarding people and things is online. Without conscious awareness, their own internal process has conformed to this way of being in the world, and they often find themselves feeling shocked by the levels of harshness, criticism, and aggression in their own responses to the world and to themselves. The bottom line is that we are influenced by what we expose ourselves to. Ultimately, excessive exposure to a world with constant judgment, evaluation, commentary, and comparison can make any of us lean toward relationally aggressive ways of encountering others and, actually, ourselves.

SELF-CENTRICITY AND ALL THE ROOMS IN THE HOUSE

At the same time, we have access to connection with a wider range of people online, we are in constant interaction within curated platforms that direct our behaviors and cater to our preferences. Our social media platforms suggest "liking" or "following" people based on shared connections or "markers" in our algorithms, and our apps request access to our contact lists in order to increase our "friend" pools. As I'll discuss in chapter 6, the self-centricity of information that comes our way because of algorithms birthed from our tech use can subtly keep us interacting with people with whom our history indicates we'll "fit." This is not inherently bad. It is, however, a threat to the diversity of our relational lives and, therefore, to the health of our being.

Just as objectification and relational shopping can lead to unidimensional and inauthentic social and relational lives, the way we comply with our devices' suggestions (e.g., suggested friends to follow or like) has the power to make us more empathic, well-rounded, openhearted people—or to close us off to these traits altogether. Think about it: Prior to the invention of the printing press, the average human saw approximately five hundred faces in his or her lifetime. Those five hundred people likely existed within relatively close physical proximity and geography of the person seeing them, suggesting at least some shared lifestyle values and traits. Given the lack of access to easy transportation, people had little choice with whom they interacted in their daily lives. In direct opposition to this reality, we now have unending choice and control over whom we interact with and how we carry out the interactions. Not only can we choose to avoid encounters with people who stretch us and make us think, but our social networks also suggest new connections that will buttress what we already like, believe, or support.

To understand the significance of this impact of technology on our relational lives, let's consider this analogy: When we are building houses, we need to know what types of rooms are functionally important and what rooms exist primarily based on our preferences. For a cook, a kitchen and dining room are essential. For a writer, a study or office may be key. For many, a comfortable bedroom may take priority. Even though it may not be considered a priority, it's unlikely any of these individuals would build a house without a bathroom. For a home to be functional and comfortable, common sense suggests that certain rooms need to be included.

The same is true of our "relational houses." While we may have personal preferences that preclude our spending greater amounts of time developing certain relational "rooms," we need to tend to the overall structure if we hope to be content over time. Granted, the relationships making up our "social" room may be filled with those developed and maintained in socially networked spaces, and that is fine. But other rooms, with the capacity for a different kind of depth and different experiences, are also important to tend to for our long-term relational health and function. We all need rooms for one-on-one communication, silence and contemplation, joyful rowdiness in groups, physical activity, and more. Only with a diversity of opportunities for embodied engagement will our full humanity unfold.

Parents would do well to consider this concept as it relates to their children. As much of entertainment, reading, learning, and even household management gravitates to digital domains, it would be easy to passively encourage a child to build a house with no rooms for silence, boredom, or deep critical thought. It's easier to place all learning on the tablet and all entertainment there too. It involves fewer arguments to let children play whatever games they want with no limits or guidelines. Parents get more done when their children are otherwise occupied, and things feel better when we aren't arguing over online content or use of devices. When we choose not to set boundaries and guidelines and we take the easy road to qualm any strife, however, the developmental "house" we build is at risk of serving children poorly throughout their lives. Setting difficult limits and inconveniencing oneself in the service of finding high-quality content free of violence, inappropriate sexual imagery, and monetization will pay off when parents notice that their children's emotional, intellectual, and behavioral "homes" have all the rooms they need for their entire life, not just those rooms they think they need now.

TAKE ACTION

Assessing the Heterogeneity of Your Relational Life

- With your cell phone or address book in hand, make a chart on a piece of paper, listing the names of your contacts in the first column.
- Across the top of the paper, write and make a column for each of these words: Gender, Race, Religious Affiliation, Political Leaning, Education Level, Living Situation (e.g., owns a home; rents a home, apartment, room).
- Then begin filling in the columns of the chart for each of your contacts.

For most people, the most checkmarks land in columns with demographics that match their own. If this is true for you, see this as an invitation to seek out some new people and relationships to stretch you and provide you with opportunities to learn and become more flexible and open in your responses to those different from you.

RELATIONAL AGGRESSION IS A (DESTRUCTIVE) REALITY

In his essay "Why We're Losing the Internet to the Culture of Hate," *Los Angeles Times* and *Time* magazine journalist Joel Stein effectively sums up the online trend toward relational aggression when he says, "[T]he Internet's personality has changed. Once it was a geek with lofty ideals about the free flow of information. Now, if you need help improving your upload speeds, the Web is eager to help with technical details. But if you tell it you're struggling with depression, it will try to goad you into killing yourself. Psychologists call this the 'online disinhibition effect,' in which factors like anonymity, invisibility, a lack of authority, and not communicating in real time strip away the mores society spent millennia building. And it's seeping from our smartphones into every aspect of our lives."

Rider University's John Suler further illuminates this effect when he makes the distinction between benign and toxic disinhibition. The former occurs when the anonymity of the online environment affords a person the ability to practice pro-social or healthy behaviors. The latter is in play when the felt invisibility of online spaces allows a user to practice behaviors that could hurt oneself or others. Relational aggression, using one's connections with others to hurt them, is one such expression and is rampant in online spaces. Follow the transcripts of conversations between clan members in video game play, read the comments section of just about any editorial or review, or simply peruse the wall of a frequent Facebook user and you'll find ample examples of sarcastic, mean-spirited, and at times downright threatening language.

Fueled by the objectification of those we encounter in digital domains and the perceived anonymity of such spaces, online relational aggression appears to serve two purposes: (1) openly criticizing or cutting people down to size and (2) having fun and getting a charge from doing so. Regarding this twofold motivation, Stein says, "The people who relish this online freedom are called 'trolls,' a term that originally came from a fishing method online thieves use to find victims. It quickly morphed to refer to the monsters who hide in darkness and threaten people. Internet trolls have a manifesto of sorts, which states they are doing it for the 'lulz,' or laughs. What trolls do for the lulz ranges from clever pranks to harassment to violent threats. There's also doxxing—publishing personal data, such as Social Security numbers and bank accounts—and swatting, calling in an emergency to a victim's house so the SWAT team busts in. When victims do not experience lulz, trolls tell them they have no sense of humor. Trolls are turning social media and comment boards into a giant locker room in a teen movie, with towel-snapping racial epithets and misogyny."

Internet trolls who sow discontent online by mistreating others, inciting disagreements, and systematically discrediting people are now nearly mainstream. These same individuals show high scores in narcissism, psychopathy, Machiavellianism, and, especially, sadism, as reported in a 2014 study published in the psychology journal *Personality and Individual Differences*. Given the complex mix of traits trolls live with, it is not surprising that online spaces have become ideal playgrounds for them.

But it doesn't require a troll to harass someone anymore. Today's most common forms of online harassment are carried out within everyday interactions and involve people preying with cruel intent regarding personal or physical characteristics, political views, gender, physical appearance, or race. Research from the Pew Research Center suggests that nearly 70 percent of eighteen- to twenty-four-year-olds and 41 percent of adults have experienced online harassment. In fact, 61 percent of the survey respondents see online harassment as a "major issue." How sad this is! We are living in a time when a majority of individuals feel impacted by harassment, yet those perpetuating the abuse have more and more platforms from which to act. In addition, the reduction in inhibition created by the distance the internet provides between the perpetrator and the target makes aggression and harassment feel divorced from any consequences to anyone other than the victims.

Whitney Phillips, professor at Mercer University and author of *This Is Why We Can't Have Nice Things: Mapping the Relationship between Online Trolling and Mainstream Culture*, rightly reminds us that while trolls are often portrayed as aberrational, the kind of harassment they perpetuate is carried out by normal people who get a charge from their edgy behavior. In fact, if we think of online relational aggression as being the domain of the troll, we miss the reality that every one of us has the power and potential to be cruel, dismissive, and hurtful in our online comments and interactions. Even when our in-person communication and manners are impeccable, when engaging at a distance within digital domains, it seems we are all vulnerable to taking on the relationally aggressive tone we find in cyberspace and to becoming a bit (or a lot) less empathic and a bit (or a lot) more forceful than we (possibly) intend.

Deliberate trolls, the flippantly disrespectful online commentator, and the passively mean poster greatly impact how we communicate about ourselves online. They also achieve a broad-ranging and pervasive tenor of relational aggression that shapes both what we share and how we respond to others online. If we want a response, we incite one. If we fear criticism, we edit what we share because we've seen how others have been attacked. If we're irritated by someone's post, we don't hold back. It's often as though we lob grenades and then turn away, oblivious to their explosion. We judge, and we forget to screen those judgments before offering them to the world. We've become increasingly comfortable with offering commentary on everything. We ask for it, and we give it. Too often this feedback is unkind or out of touch at best and brutal at worst.

There are few places where relationally aggressive language is tossed around for "fun" more often than in MMORPGs (massively multiplayer online role-playing games). These games, known for their rich graphics, sophisticated soundtracks, and intense strategy requirements, offer a primary place of community for those who play them. Joining a team, an individual interacts with players around the globe, talking with them via microphones and listening with headphones. For gamers who engage these environments, the clans and communities they create within the game are often the people to whom they feel the closest sense of attachment. This is tricky because these environments are also built to stimulate all the players' senses, increasing a strong competitive vibe and driving players to move up in the community standings.

Given the content of many of these games, the kind of verbally aggressive banter between both teammates/clan members and opponents is intense. As discussed in the previous chapter, we do well to remember that individuals who are immersed in these environments have adrenaline surging, triggering an emotional response that is poorly modulated and most likely out of step with the actual physical environment in which the individual abides. This dysregulation is occurring at the same time the gamers are hyperfocused on the task at hand. Add to this language that is dismissive at best and harshly critical and damning at worst, and you have the perfect storm for a participant to react in kind or to retreat in shame. This sets up a relational matrix wherein individuals repeatedly experience relational aggression as a normative and acceptable way of treating people.

For women who game, the reality of aggressive treatment is especially real. The issues here are complex and varied. Even as the demographics of gaming are changing and women constitute a significantly larger portion of the community of creators and consumers than in the past, gaming spaces are often rife with harassment of women. Any of us who are connected to women who spend time in these spaces would be wise to offer ourselves as listeners and comforters to those who suffer appalling misogynistic slurs, insults, even threats during play time. We also should be paying close attention to the plethora of popular games that allow and perpetuate horribly sexist stereotypes that encourage sexual assault within play and that generally demean those who identify as female. Letters to the creators and producers of such games and conversations with the players are important if we are to create a more welcome online world for women.

Note that not all video games are violent and misogynistic, and that not all players fall into aggressive patterns. We are complicit, however, in the passive normalization of the relationally aggressive interpersonal dynamic, including mistreatment of women, when we do not take seriously how the places where we spend time impact us and how they shape the way in which we relate to others.

TAKE ACTION

Take time to consider how people are treated in the digital domains you inhabit, and make changes if needed. Some questions and responses to consider include:

- Are the platforms and apps you frequent built in such a way that people are treated as commodities? Are people objectified based upon their gender, appearance, socioeconomic status, or race? Are individuals ridiculed and judged openly? If so, consider reducing or avoiding engagement in these spaces and educating others about the realities of discrimination and mistreatment that exist within them.

- Consider the positive reasons you engage the space and see if you can find other digital domains that meet these needs without harming others. For example, if violent video games are your "vice," consider what it is about the game that draws you in. Is it the rich visuals and pounding audio tracks? Is it the strategy involved in the game? Is it the sense of community you find with those with whom you play? Is it the competitive angle you like? Once you determine your primary draw, see if you can find a game that engages those elements but does not inoculate you against what might be considered "normal" responses to violence and misogyny.

RELATIONSHIPS AS PERFORMANCE ART

As more and more online spaces develop opportunities for users to broadcast in real time, the issue of relationship development and maintenance as "performance art" also becomes real. Treating one's life or hobbies as the subjects of live broadcasted experiences often means that a person is working to amass a following in order to gain either status or income. This is true in spaces such as Facebook Live, where people can livestream whatever activity they are doing, and in Twitch, where gamers set up channels to stream videos of themselves playing video games while also interacting with viewers during the stream. On Twitch, once players have created a channel, they strive to build a high number of subscribers by creating streams that are interesting enough to hold the attention of viewers for long periods of time. Individuals sign in to their channel, where they see a split screen with one part broadcasting a video of the player and the other displaying a live feed of the in-game action. During their live broadcasts, players keep tabs on who is watching and will do shout-outs to them, making viewers feel part of the action. To amass the necessary quantity of subscribers in these platforms, players must be charismatic or have unique skills and capabilities. For those who fit these categories, success can mean a significant income (some well-known Twitch broadcasters are making upwards of two hundred thousand dollars per year), as well as the personal accomplishment of having and maintaining a significant following. For those who don't possess either the skill or charisma to succeed in this way, real and intense feelings of failure can abound.

Just as performance art exists to challenge the preconceived ideas around visual art, these new digital ways of creating relationships between individuals and a community are challenging our traditional ideas about relationships. Do connections built on display by one party and consumption by another fit the definition of "relationship"? For a connection to be meaningful, does it need to include back-and-forth communication? Is our worth and value tied to the quality of our online presence or performance?

In a smaller way, even our quite mundane relational interactions now occur in public spaces that serve as venues for amplification. We respond to someone's Facebook post about a meaningful happening rather than doing so in a private channel, realizing that others will read our response. We post our greatest successes widely while failing to mention the more mundane or difficult parts of our lives, knowing that this creates a certain image of ourselves and our lives.

So much of what has historically been developed in relatively private spaces now happens in front of audiences large and small. This leads to my counseling everyone: we must be mindful of the digital "documentary" we are creating, just as we must consider how living online affects our tendency to initiate, build, and maintain friendships. In short, we must pay heed to the authenticity, or lack thereof, that drives our relational selves.

TAKE ACTION

Align your social networks with your embodied, relational ones.

- Do an inventory of your social networks (whether they are via video games, on platforms like Facebook, etc.).
- Consider whether you are engaging with people on your social networks who are there only for you to show off to, or others who lead you to feel "less than."
- Assess the newsletters and online subscriptions you receive.

Then carefully consider who and what are positive influences in your life that you want to continue connecting with, and who and what might be best to part ways with. This need not be a harsh rejection session but rather a realignment of sorts.

THE WISDOM OF THE OTHER

As increasing amounts of our interactions occur in writing or recorded voice memos to be read or listened to in isolation, fewer of our interactions occur within the context of a community that might help us make sense of relational realities and provide feedback on our communications. When we talked at the dinner table or on the phone connected to the wall in the middle of the house, others in our proximity could offer insights on how we might become better communicators. When we had discussions in class (as opposed to online), we learned in actual time, by trial and error, how to communicate what was on our mind and how to be part of something bigger than ourselves. We learned how to have fruitful and neighborly discourse on the spot.

As these more embodied experiences move to online spaces where words are typed and where people can tap out thoughts with little to no consequences, it will require more intentional work on all our parts to make sure we keep our relational skills in tune. To this end, ensuring that at least some of our communication occurs in settings where we can benefit from feedback from trusted others would be wise.

MORE IDEAS FOR CREATING HEALTHIER RELATIONSHIPS OFF- AND ONLINE

EXAMINE YOUR RESPONSE PATTERN TO ONLINE INFORMATION SHARING BY OTHERS.

When you learn something from a post that someone shares online, take a moment or two to consider how it might feel to respond to that person alone rather than to the entire network. See what emerges.

TAKE A TEN-MINUTE PAUSE BEFORE POSTING OR RESPONDING TO POTENTIALLY PROVOCATIVE INFORMATION.

When you feel particularly stirred by a post you find on a social network, take note of your feelings. Explore those feelings in your own mind. Rant (out loud if need be) in your own space. Once you have taken time to explore your thoughts and express your feelings, see if you can construct a response that is appropriately respectful of the humanity of the person to whom you are responding.

PRACTICE NONJUDGMENTAL AWARENESS AND RESPONSIVENESS.

Most of our thoughts about others are driven by our automatic judgments of them. Commit an hour or a day to living by the motto, "Be kind to everyone, for theirs is a difficult journey." See how leading with empathy and openheartedness changes the tendency toward judgment and categorization.

SOMETIMES, PICK UP THE PHONE OR INITIATE A VIDEO CHAT.

Every once in a while, forgo the impulse to text and make a call instead. Don't let yourself off the hook by thinking, "I don't have time for a long conversation." Instead, allow yourself to say to the person you are calling, "I only have a minute or two but thought it would be a nice change of pace to actually speak voice to voice." Don't be afraid to end the call when you need to.

LEAVE YOUR PHONE IN THE CAR WHEN MEETING WITH OTHERS, AND DON'T WEAR EARBUDS WHEN OTHERS ARE PRESENT (AT LEAST SOME OF THE TIME).

This one needs no explanation. Just do it.

WAIT IN LINE, AT A MEETING, OR ELSEWHERE WITHOUT INTERACTING WITH YOUR PHONE.

Resist the feeling that you will look silly or that it will be awkward if you catch someone's eye. Use this time to actively seek out eye contact and practice it or to experiment with small talk. Both skills are on the decline and are super powers in today's economy.

PRACTICE EYE CONTACT.

Find a friend who is willing to practice with you the uncomfortable work of looking into each other's eyes. Try to maintain eye contact for three to five minutes without talking. You'll feel silly and squirmy at first, but over time and with practice, you will find this to be an incredibly rich experience. Don't be surprised if this experiment increases feelings of connection with this person.

HANDWRITE A LETTER OR NOTE.

Taking the time to write by hand can force us to be more thoughtful about what we share. Receiving such a letter imparts feelings of meaning and worth. Make this easy for yourself by keeping pre-stamped envelopes and cards in your backpack or desk. Set an alarm that reminds you to send one card per week.

PRACTICE FINDING THE GOOD IN OTHERS, AND PERIODICALLY AFFIRM SOMEONE IN PERSON.

Take some time to practice praising people. When you encounter someone, try to find at least one thing about that person to affirm. When you feel irritated by someone, work diligently to find even one small trait in that person you could, if given the opportunity, mention.

6

Technology and the Self

The "self" is an expansive topic that has been considered throughout history and referred to by names such as "soul," "psyche," and "essential identity." Despite conflicting theories regarding the development of this foundational element of our humanity, some generally held ideas about what it means to live with a *cohesive and stable sense of self* do exist. At root, people with such an identity are able to:

- Experience their self as distinct from others.
- Connect to and separate from others in healthy ways (being neither overly dependent nor overly independent).
- Have a general sense of core values and worldviews.
- Perform general processes related to being active participants in the world.
- Handle consequences related to their actions in the world.

Note that mobility or cognitive capability is less a part of a cohesive and able self than are the deeply intrapsychic activities of knowing oneself as a unique being within the context of relatedness to others and living that out in whatever capacities are possible.

THE DEVELOPMENT OF SELF

For the sake of a shared vocabulary, I propose the following oversimplified process to describe the development of a self:

1. A baby is born.
2. The baby does not have the physical, emotional, or cognitive skills to care for his or her needs independently.
3. Parents or other caregivers attend to the baby, watching and responding to cues the baby presents. For example, when the baby cries,

a loving caregiver works to discern the cause of discontent by checking diapers, offering comfort items, and determining if hunger might be the cause. When caregivers ascertain the cause of the baby's discomfort, they take actions to resolve it. In other words, they learn to attune to the baby and to read the baby's needs, with the goal of providing embodied connection, soothing, and security.

4. Over time, children learn, *by experiencing the pattern of discernment and care offered by caregivers,* how to understand and address their own impulses, needs, and wishes. Eventually, they begin to make efforts to care for themselves. In other words, by experiencing and observing (consciously and unconsciously) how caregivers understand and minister to their needs, children begin to see a path toward performing these functions for themselves.

5. In fits and starts and through experimentation and copying behaviors (the caregivers' actions), children learn to provide primary care for their own needs as well as learn to notice and respond to others in their context. Ideally, children develop an ability to both connect to and separate from others in healthy ways, being neither overly dependent nor overly independent. They learn to know that their thoughts and feelings are natural parts of life that can be addressed and altered or embraced.

6. As time goes on and a child grows in cognitive and emotional capacity and understanding, worldviews, beliefs, and core values begin to surface. With maturity, these come to create a foundation from which the growing or grown human functions.

7. As development continues, the person can live either from a conscious awareness of this foundation or from a primarily instinctive (outside of consciousness) gut reactivity. A more cohesive sense of self grows from at least a partially conscious awareness of one's distinctness from others. The more people are aware of their basic values, worldviews, biases, temperament, emotional range, and cognitive style and proclivities, the more likely they are to experience themselves as "grounded" and capable of caring for themselves in a complicated and diverse world. These are manifestations of a solid sense of self. These individuals can find their way to emotionally regulated states when charged situations arise, even if it takes them a while to do so. They can maintain relationships with diverse people. They do not, on average, function in enmeshed or enabling ways and can attach and separate with flexibility. Although they are impacted by others and by their own internal states, they can access a sturdy foundation within themselves from which to act, react, and respond to life.

8. Disruptions to any part of this process can result in what the psychological community calls "developmental arrests." These function

as "stuck spots" that hold people back from further development. All manner of efforts are made by those experiencing a developmental arrest to either compensate or resolve the unmet needs that hold them back. Compensatory behaviors or ways of being buttress them as they continue to grow, but specific parts of their development may stay underdeveloped until the original and basic needs are addressed and met.

This is of course a grossly oversimplified recounting of a deeply complex process. Few things about the development of the self are simple, and no one's development is perfect. I have yet to meet someone who has moved through the process unscathed, and in reality it is in the scathing process that some of the hardiest selves are developed. In fact, it is in the learning to handle the less-than-perfect experiences that much of our character building takes place.

Perhaps the most important concept for our discussion here is the vital role that caregivers play in the development of the self. When the providers of care in a child's life see the precious uniqueness of the individual for whom they are caring, help the child see the same, and encourage the child to move from dependence to independence (and ultimately interdependence), then optimal development occurs. This is also the part of development fraught with the most complexity. All humans are raised by other humans who have their own biases, beliefs, and preferences. Caregivers are not immune from this. Often they provide care more from their own assumptions and resulting toolkits than by relying on the cues of their children to inform them. Ideally, a caregiver would provide care that is specific to each child's unique abilities and that addresses the fostering of skills the particular and unique child may need. Unfortunately, this does not happen all the time. Even when they do the best they can, parents are limited by their own knowledge and skills and at times have a difficult time parenting flexibly.

For example, if a particular child is prone to being soothed by silence and stillness but is parented by a caregiver who sings loudly and bounces, that child's deeply internal need for silence and stillness may stay present but unmet. A child who sees the world through the lens of structure and pattern but who is raised by a caregiver who values flexibility and spontaneity may arrive at adolescence with an under-addressed need for order. These kinds of mismatches and resulting developmental arrests happen more often than not.

But even in a world where these mismatches are inevitable, it is possible to be a "good-enough" parent—a caregiver aware of personal preferences and biases, attuned to the unique needs and personality of the child, flexible and open-minded in communication and action, and humbly aware of one's humanity. People who are fortunate enough to receive good-enough care or who have been offered the ability to learn and develop themselves in well-rounded ways will ideally experience a cohesiveness of their self at the core. These individuals:

- Demonstrate an ability to know and activate their preferences and styles.
- Evaluate their goodness of fit or lack thereof in whatever setting they find themselves.
- Experience a range of emotions and have an ability to regulate them.
- Can connect to and separate from others without fear of losing themselves.
- Adjust and flex without compromising their values.
- Live life with intentionality.

In general, those who have mastered even a rudimentary self-structure sense a constant place of security, even if it takes deliberate work to find it. At their core, they know they are worthy and valuable in and of themselves, and they can access a place of groundedness at their center. Think of this solid center as the time signature in a piece of music. If we listen closely to any given song, we will likely hear a deeply embedded rhythm being held by a percussive instrument of some kind. This basic rhythm holds the disparate players of the music together in an integrated way. When we can identify some of the core qualities, values, and tendencies of our authentic self, we can use these as our integrating rhythm, adding other skills, abilities, and experiences in keeping with the core of who we are. But we can only reach this level of self-development when know and trust ourselves at the core.

When an individual arrives at adolescence or adulthood without having developed a cohesive sense of self, the world can feel like a threatening place. Without access to a personal inner place of grounded assuredness, a person is left looking to the outside world for comfort, entertainment, information, and motivation. Although I'll delve more deeply in a later chapter into this state of being when I talk about having an "external locus of control," what to know now is that individuals who function from this outside-in way of being are dependent on the external world in highly vulnerable ways. When they face hardship or even glee, they look to others to see how to respond. When they experience difficulty, they look to factors outside themselves to blame, or they interpret shame coming toward them from all external sources. Emotional regulation is very difficult for these individuals, and the ability to be still or meaningfully with one's self is nearly impossible.

Research presented in Jean Twenge's groundbreaking book *iGen* exposes themes important to our discussion here. Focusing on generational differences, Twenge finds that while members of Generation X stretched adolescence beyond previous limits, millennials and those she refers to as "iGen" (born between 1995 and 2012) are contracting it. Consider that adolescents today are under their parents' roofs more than ever but are interacting with them less; that they are less likely to pursue their driver's licenses than their counterparts of other generations (until pushed to do so); that they are drinking, dating, and working less; and that they are spending less time unsupervised. The confluence of these realities, according to Twenge, suggests that childhood now stretches well into high school. She finds eighteen-year-olds now acting much more like fifteen-year-olds used to act and fifteen-year-olds acting much more like thirteen-year-olds used to. This results in a sort of altered developmental norm within present culture and appears to leave adolescents more prone to depression and suicide. All of these realities greatly impact the development of a sense of self. One can't help wondering if (assume that) the time our millennials and iGen kids are spending with technology isn't dramatically affecting their developmental trajectory. Perhaps hanging out on devices in the comfort of home is just too appealing and easy, and disinclined to bring about the kind of life that develops a cohesive and deep sense of self.

THE IMPORTANCE OF SILENCE AND TECHNOLOGY AS A USURPER OF IT

One of the most reliable ways to develop and nurture a sense of self is to foster an ability to be with one's self in relative silence. Cultural traditions, rites of passage, and coming of age rituals often incorporate such times. Therapy, faith journeys, and mindfulness practices do the same. Regardless of the platform, silence and the boredom or openness it fosters are crucial avenues for the development of self.

Before our world became as hyperconnected as it is today, having space and time for quiet was common. Children played in relative silence, using only their imaginations and whatever resources they had available to entertain and occupy them. We received calls and mail in set and determined locations. The broadcast world went silent at night. Without romanticizing the past, we must admit that we live in a time when naturally occurring opportunities for stillness and lack of stimulation are rare. This does not mean that all individuals who were raised when life moved more slowly and spaciously automatically developed a whole and hardy sense of self. Plenty of opportunities for arrested development and compensatory dependence on things outside one's self existed then as well; there were just fewer of them, and they weren't as easily accessible.

In today's economy, the opportunities for silence and spaciousness are few and far between. In fact, rooms that historically have been seen as those for rest and self-care, the bedroom and bathroom, are now places where many people engage their technology most heartily. Taking one's tablet to bed to read is commonplace, as is checking the news and one's social media feeds upon waking, long before getting out of bed. Tweets are fired off while sitting on the toilet, given that it is one of the only places remaining where we are still long enough to compose original thoughts. Even the very young are not immune to this new trend. Consider the iPotty, a tool/toy designed for children learning to use the toilet. This training toilet includes an "arm" that extends to hold a digital tablet that children can watch or play games on while they take care of their bodily needs of elimination. Chett, a resident director at a college where I spoke, suggested we start a new movement (get the pun?). The crusade would encourage people to use their time in the bathroom as practice for tolerating boredom and could be called "Shit Still." I think the idea is brilliant!

Along these same lines, screens are popping up everywhere. Hotel restrooms are now embedding digital screens into the mirrors above sinks that showcase music videos or a lifestyle news channel, gas stations are featuring on-the-pump television screens that blare lifestyle content or the day's news, and grocery store checkouts are streaming cooking shows on monitors dangling over the counter where we unload our carts. A restaurant I recently visited had an individual screen at each booth with options of sports and entertainment content from which to choose. Even churches have entered the game, broadcasting sermons from one location to multiple others via large screens and constantly redirecting the focus of attendees to screens in all parts of the service.

Think about it: our everyday embodied spaces are now designed to maximize our comfort with and attention toward screens and devices. Ergonomic seating for interaction with keyboards, charging stations, and screen-friendly lighting are primary considerations for interior design specialists, eclipsing a focus on designing for embodied interaction. Consider that windows, which let in natural light and open our minds to the world outside ourselves, are now thought to cause problems for viewing our screens. In many high-tech workplaces, people are constantly closing the blinds to avoid glare on their screens, ultimately closing themselves off to the natural world around them.

This is a serious problem, because expansive connection with the environment is one of the primary ways we can find the kind of silence and spaciousness that helps us develop our sense of self throughout our days (specifically) and lives (generally). The antithesis of overstimulation and distraction, the natural world—from the sky to patches of grass—is filled with opportunities for visual and auditory silence. In fact, the backdrop of silence in nature, as opposed to the ambient backdrop of noise in our everyday lives, can be aspirational as well as inspirational. Noticing the ways that our contexts are changing in ways that limit our engagement with the therapeutic powers of nature, silence, and spaciousness is critical to well-being.

TAKE ACTION

Ideas for Creating and Using Silence

SET SOME TIMES OF THE DAY FOR TURNING OFF ALL ELECTRONIC SOUNDS.

Even if you feel you work less efficiently in silence, try it at least some of the time. If you typically listen to podcasts while you do physical tasks or to music while you do mental ones, take the opportunity to work for at least fifteen to thirty minutes with no sound at all. If you are new to silence, you may need to work up to this in increments, beginning with five minutes.

PRACTICE A GUIDED MEDITATION, THEN TRY IT ON YOUR OWN.

After doing a guided meditation or two (easy to find free ones online at various sites such as MARC.UCLA.edu), try your hand at doing your own with no vocal lead. Even if you try to sit in one place and breathe in silence for just three minutes, it counts! Do this with frequency.

USE A SINGING BOWL.

Tibetan singing bowls are great tools for becoming acquainted with silence. They can be purchased in Tibetan shops or online at nearly all price points. Resting the bowl on the flattest part of your hand, being careful not to have your palm or fingers touching the sides of the bowl, use the striker or ringing stick against the rim of the bowl. Feel the vibrations of the sound while you listen to the quality and layers of the sound itself. As the sound and the vibrations diminish, let the fading stimuli lead you physically into silence. Once all ringing and vibration has stopped, see if you can listen to the silence in the same way you listened to the bowl.

TECHNOLOGY AS A DISRUPTOR IN THE DEVELOPMENT OF SELF

Just like it colors the spaces we inhabit, technology also has the potential to enhance how we develop. Information, ideas, people, places, and experiences that can shape us are available to us in ways they have never been before. When we engage digital spaces to the end of becoming wiser, more well-rounded people, the potential is nearly limitless. But the likelihood of this kind of positive engagement with technology is predicated on consumers being discerning, self-aware, and intentionally engaged enough to determine what content and how much of it will enhance the development of their self and what might hinder it. This type of vigilant use is lost on most of us much of the time.

With so much of our experience being influenced by screens that feed us content indiscriminately, practicing care about what we take in and how it might affect us can be challenging. With screens engaging us while we pump gas, wash our hands, pray in church, and even meditate, we are being inundated with consistent, subtle reinforcement of the idea that we are, in and of ourselves, not enough. We can begin to believe to our core that we must have input from others, that more information is better, and that there is little merit to the silence that might exist if we were with our own self in consistent and meaningful ways.

Perhaps most disruptive to the development of self are the smartphones with which we wait in line, drive, and even sleep. Not only do these provide us an escape from ever needing to be with our own selves apart from external stimulation, but they also offer a place for us to offload our own choices. We no longer need to decide where and what we want to eat; we let Yelp decide for us. We don't need to budget our time in order to accomplish what needs to be done; we just use apps and delivery services. When we can't fall asleep, we simply scroll through our feeds. This constant presence of technology contributes to a sense of constant potentiality and creates a background noise of sorts—again, sending us the subtle message that we, ourselves, are not enough to meet our own needs. All the while, this keeps us from developing the hardiness of self that could prove otherwise.

Since 1975, the Monitoring the Future Survey, funded by the National Institute on Drug Abuse, has been assessing what's happening with American adolescents. Through more than one thousand questions, the survey consistently has found that subjects who spend more time than average with screens are more likely to be unhappy and that those who spend less time are more likely to be happy. In fact, eighth graders who engage social media six to nine hours per week, an average amount given current statistics, are 47 percent more likely to say they are unhappy than those who use social media less. While happiness isn't necessarily an indicator of a strong sense of self, the ability to feel content, at the least, is important for the development of a grounded self. It seems clear that engagement with screens may counter this ability.

Finally, many of the apps with which we interact actually interrupt the kind of feedback loops that lead to a healthy development of self. As noted earlier, as humans we learn to care for ourselves by initiating feedback loops that enable us to master the basic skills of living. A baby is wet and therefore uncomfortable. He cries. His parent changes his diaper and interacts with him lovingly, providing feedback about how to resolve his discomfort. A child is bullied, resulting in feelings of fear, sadness, and anger. She acts out of sorts and the attentive caregiver engages her to work through her feelings, providing feedback about how to know and work through the situation. An adolescent dislikes his appearance. An attentive other notices subtle cues and creates opportunities to talk with him and help him find ways of working through his strong feelings. This might include finding active ways of building confidence or of appreciating parts of his physical being or could include identifying realistic and meaningful ways of working to alter his body with exercise or his appearance via clothing choices.

With a healthy focus on the child and on ways he or she can live and work with discomforts or even strong positive feelings, these feedback loops direct the attention back to the needs of the core self and build the skills of the person to meet those needs. Rather than fostering a dependence on the caregiver, these feedback loops help the child look inside, explore emotions, and come up with ways of working through them.

Social networks and other new media have the potential to significantly alter these feedback loops. Instead of relying on trusted, wise others who will direct us back to our own internal compass to determine how to feel about ourselves and the world, we now look to external reinforcement and feedback to shape our very selves. At the time this book was published, several former Facebook executives had publicly expressed regret over the way in which the social media giant had provided these kinds of self-centered but not self-grounding feedback loops. When we stop to consider that we now live in a global economy—where people engage social networks to garner feedback, input, comfort, and reinforcement rather than to contribute as whole selves to the embodied world—the consequences are huge.

Chamath Palihapitiya worked for Facebook between 2007 and 2011. In the early days, he claims, not much thought was given to any potential negative aspects of the development of the mammoth social network, and he speaks of "tremendous guilt" about what he perceives as an erosion of the kind of common decency that grows from being well-grounded humans. He notes, "We curate our lives around this perceived sense of perfection, because we get rewarded in these short-term signals—hearts, likes, thumbs up—and we conflate that with value and we conflate it with truth. And, instead, what it is, is fake, brittle popularity that's short-term and leaves you even more, admit it, vacant and empty before you did it. . . . Think about that, compounded by two billion people." He goes on: "It literally is a point now where I think we have created tools that are ripping apart the social fabric of how society works. That is truly where we are. . . . The short-term, dopamine-driven feedback loops that we have created are destroying how society works: no civil discourse, no coopera- tion, misinformation, mistruth."

When we live to receive external feedback, we shortchange the process of learning to function from our actual core. This leaves us dependent on others and lacking in a grounded sense of self in truly unhealthy ways.

"STUFF" AS A THREAT TO THE DEVELOPMENT OF SELF

Just as our exposure to a smaller pool of faces and people may have allowed for more metered opportunities for self-development in the past, our access to "things" also impacts the nature of our core selves. One of the first ways we experiment with the ownership and stewardship of things is through play. It is widely believed that the nature of play changed in 1955 when the Walt Disney Company sold its first television advertisement slot outside the holiday season. The advertisement was for the Mattel Thunder Burp Gun and aired during the *Mickey Mouse Club*. Prior to this, people lived by the long-held idea that play was largely dependent on people utilizing their imagination and whatever objects they could get their hands on. The rare toy was considered a very special object to be treated with care. With the advent of mass production and advertisements for toys outside the holiday gift-giving season, a new way of thinking about play began to take hold. Quickly, play became less about the process of creation and more about the utilization of objects. When a Thunder Burp could be had, how did a gun-shaped twig hope to suffice?

The continuation of this desire for objects over creative invention has been buttressed by increasingly sophisticated marketing and by the avail- ability of a range of toys at all price points and quality levels. In fact, these patterns exist in all aspects of our lives, with the accompanying reality being that the internet can deliver endless access to unlimited items for us to amass. There's something for every purpose and no need to consolidate or share; there's plenty for everyone to have everything! When we do face a situation in which we need to employ creative thinking or invention, we have easy access to limitless online avenues for instruction, allowing us to sidestep the process of engaging in the kind of creative problem-solving skills and resourcefulness that contribute to the knowledge that we can rely on ourselves and that we are capable of finding solutions. Getting to experience and develop, in real time, our abilities to think critically and creatively goes a long way toward helping us feel solid in our core.

The subtle and not-so-subtle message that now begins in childhood and extends throughout adulthood is that we can easily acquire both objects and states of being outside ourselves and that those things are better than what we could resourcefully make or achieve on our own. There are certainly challenges to this claim, which we see in the maker or simplicity movements that redirect us to the value of developing a sense of resourcefulness about that which we desire and own. But unless we actively engage these countercultural movements, the message with which we are constantly bombarded, that "stuff" will make us content, is so powerful that it can insidiously creep into our lives. Over time, we may begin to feel that we will be okay only if we have this piece of clothing, that makeup, a growler of that beer, or this sporty car—thus helping our economy but doing nothing for ourselves. Once completely ingrained, the insidious glorification of *meaningless* stuff will keep us from having to utilize our creative and critical-thinking skills to survive in *meaningful* ways in the world.

HOW THE GAZE HELPS WITH DEVELOPMENT OF THE SELF

As noted in the "Technology and Relationships" chapter, prior to the invention of the printing press, average people saw approximately five hundred faces in their lifetime. This reality naturally limited humanity's exposure to differences and relationship potential. The fact that we can now not only see but also interact with an unlimited number of people has many implications for the development of the self. Most obviously, the increased access to an eternally wide range of people means a greater number of opportunities for growth and learning. This is profound. Having the ability to encounter people who can expose us to ideas, experiences, and opportunities we would not otherwise have is an amazing gift. The difficulty enters when we access this limitless pool of people more consistently or with more dedication than we access either embodied individuals who have proven themselves trustworthy and reliable or even our own mind and heart. Furthermore, the fact that the *history* of our online activities determines much of what we are guided to online means that if we are not careful, we end up using what *could* be a great resource for stretching ourselves as a never-ending loop of the same content. With the same-old, same-old coming before us over and over and over, we risk stalling our own development. With access to never-ending sources of influence, we risk forgoing the crucial role that consistent, good-enough others play in the development of our selves.

We must take seriously that we are, at some level, shaped by that which we gaze upon. In our earlier discussion of human development, I wrote of the caregiver discerning the needs of the child and working to meet those needs in specific and relevant ways. This process is facilitated in many ways. Gazing, or making intentional and responsive eye contact as a form of communication, is an especially powerful and essential action that caregivers take in helping children learn to know and care for themselves. To gaze at someone is not simply to look at them but instead to use the eyes to communicate a sense of both wonder and knowing. We all know, or can imagine, what it feels like to be held in a loving gaze or, conversely, in a shaming one. In the act of gazing, both the see-er and the seen are impacted. When caregivers reach past their own biases and preferences and gaze into the experience of the child for whom they are caring, important neurological and psychological processes are initiated.

If the gaze of one person to another has the power of initiating neural integration and emotional regulation, how can we say that the images we gaze at on the screens that live so closely with us do not matter? I don't believe we can. With as many opportunities as we have to gaze at a never-ending stream of highly manipulated and targeted-to-our-preferences images, it takes discipline that many of us do not have to live from a deeply personal core rather than from a place of constant scanning where we seek stimulation and soothing outside ourselves.

With our gazes distracted by so much stimulation and our opportunities to develop resourcefulness at a minimum, it is no wonder we find ourselves reaching out to technology with increased frequency. We basically bond with technology rather than with the embodied others around us who might help us grow and mature. When we do this, a cycle commences that is difficult to break: We experience something that creates a response in us. We don't feel aware enough of our own deepest needs or how to meet them, and we have underdeveloped self-regulation and creative problem-solving skills. Adrift and not knowing how to find our center, we look to our devices for the "gaze" and approval of our community, for answers, or for distractions.

Take, for instance, a disappointing interaction at a business establishment. We don't get the service we want, so we feel disappointed, taken advantage of, significantly stirred up, and uncomfortable. Instead of taking the time to address our feelings and put our minds to creative options in the moment, we turn to our phones to blow off steam. We write a blazing negative post and weigh in with a poor rating, unconsciously seeking comfort in the responses we might get or feeling as though we've taken action to resolve our mistreatment (by publicly shaming the business).

Most of us do this in various ways every day. When our feelings of fear or powerlessness are piqued by a snippet of news or a public event, rather than pulling away and looking inward to discover what we are feeling and to determine a course of action, we spend time trolling for information that will support and fuel our frustration. We then post the information or share parts of it in an effort to justify or relieve our strong internal experiences, feeling as though we have taken some kind of real action. Or we feel unfulfilled and binge on video games or streamed entertainment, gazing at the screen rather than turning our eye toward our lack of fulfillment in life in an attempt to discern more meaningful ways of getting our needs met.

In our earliest days, if we were fortunate, we received help in regulating our emotions by looking into the eyes of a caregiver who helped us know that things would resolve, that we would find relief. Over time, we internalized the steadying gaze of those who cared for us, and we were able to calm and motivate ourselves on our own. To do this work of internalization and regulation, we must at times be alone so that we can do the deep work of gazing at our core selves. What I notice more and more, however, is that when we carry a device, we have an awareness that we are never truly alone. Even when we do create time for being with ourselves in meaningful ways, we often interact with our devices in those spaces. We take pictures for later, make sure we're caught up on the news and our social networks, check for directions and reviews, or send a few texts. Might we do this because we are simply unable to hold our own attention in and of ourselves? Might we gaze at our screens because an internal gaze feels so uncomfortable and unknown? When we are not capable of turning our gaze inward, we are at risk for many compromises in the development of our true selves.

The compromises I speak of here are certainly not exhaustive, nor are they exclusively tied to our use of technology. Since the beginning of time, I assume it has been hard for humankind to muster the time, energy, and concentration necessary to foster an authentic and cohesive self-structure. With hunting and gathering and skinning beasts for dinner, time for self-reflection and intentional decision making about aligning one's values with lifestyle likely did not make the daily cut. Even still, technology offers some unique challenges to the development of the kind of self-structure to which I am referring. I believe these unique challenges lead to certain categories of difficulties that I find emerging in our culture.

To develop a strong sense of self, we must be able to function in meaningful and deliberate ways, focusing on both long-term trajectories and the actions required in the short term that will move us forward. For these actions to be possible, we must demonstrate an ability to focus, an ability to delay gratification and action, and emotional regulation. These skills are not present at birth. Developing them takes deliberate effort and practice. With a gaze trained toward technology—with its fast-moving, always-changing, highly rewarding content—that effort and practice may be challenging to come by. Without those important life skills, though, what will be lost is a strong sense of self-control, which is key to feelings of happiness and satisfaction in life.

TAKE ACTION

Research Making Eye Contact versus Looking Away

Go for a walk in a populated space. In five-minute periods, practice either looking up and into people's eyes or catching their eye and then looking away. Notice how each activity feels, both in relation to yourself and to the other. After you have done both of these, practice making eye contact and trying to communicate welcome without words, only your eyes.

TREATING THE SELF AS A COMMODITY AND THE RESULTING SELF-CENTRICITY

Not only does our tech-drenched daily life offer more opportunities for distraction, but it also creates a reality within which we are a "product" of sorts. Marshall McLuhan famously opined in 1964, "The medium is the message." Applying this thesis today, although we may unconsciously act as though we are clients of the social media, video games, and viewable/ purchasable content we engage online, we are actually their products. This reality ultimately reduces us (mostly the data we generate) to a commodity. And we pay a price as a result of this.

Not only do we tender massive amounts of helpful (and free to them) data to those companies that use us as their products, but the algorithms created by the items we click on, websites we frequent, subjects we search, and goods and services we buy online determine what pops up on our screens every day. Given that we may never consciously consider this, it is not surprising to think we might unconsciously experience these curated images and suggestions as reinforcement for our own preferences. For instance, people whose internet history is rife with searches about cars and automotive repair will be presented advertisements in the same vein. Their Netflix suggestions will begin to reflect similar themes, and their favorite news sites will feature automotive-related stories early in the feed. Rather than thinking, "My history has created this reflection of my interests," a person in this situation might easily jump to the conclusion, "My interests are in keeping with those of the world, given that the top news stories and suggestions in my queue are all about the automotive industry." While this might seem a subtle shift in consciousness, I am convinced it is a profound one.

When all we are fed is more of what we prefer, we become overly comfortable in our own little spaces. Seeing the seemingly serendipitous nature of our online experiences as being evidence of how perfectly aligned our own preferences are with that of the world keeps us in a self-centric loop. We must recognize that this is occurring and that it does so because, ultimately, the corporations that collect and control our data are benefited by us having positive feelings and experiences online and of staying on their sites for significant amounts of time. For many of us, this means using our data to feed us more and more of what we already like and to keep us from being in uncomfortable spaces. The self-centricity that unconsciously comes into play blinds us to the ways in which we have become closed-minded or narrow in our experience. When we remove the blinders, we are faced with extreme all-or-none, right-or-wrong, my-way-or-the highway kinds of mind-sets.

We all have been exposed to the stereotype of recluses who stay tucked away in their own mental and physical environs, becoming increasingly paranoid and certain they are the only sane people in the world. Sadly, our devices provide impetus for all of us to become recluses in some form or another. The more we interact with our devices, the more they begin to reflect our own little worlds back to us. We only eat at places for which we can find reviews that fit our liking. We only read articles that are suggested to us based on our online histories. We stalk people to find out if they agree with us or are like us, and we only relate to them if they are. We watch previews and read headlines and only delve into the actual content if we decide ahead of time that we'll appreciate it. All these things keep us from facing the kinds of diverse experiences and realities that might make us more well-rounded and flexibly adaptable people—as well as better neighbors and citizens.

In some ways, even the act of listening to music used to help us become more well-rounded people. When radio or long-play records were available as our primary means of exposure to music, we were confronted with songs we were unfamiliar with or that were regularly outside our known setlist. Omitting unfamiliar or disliked tracks was difficult or largely impossible, so we were passively exposed to newness and novelty. This inherently had the effect of growing and stretching us as listeners. With the advent of music selection sites, we can now assure ourselves of a constant stream of music that is exactly what we like. And we are increasingly comfortable with removing the "inconvenience" of listening to something out of our ordinary.

This kind of curation happens in many ways across many domains of our lives and is not necessarily a bad or negative thing. It is, however, not a completely neutral or benign reality. We are healthiest when we face optimal levels of challenge to our automatic ideas and beliefs. When these are never confronted, we can remain blind to our own biases or, worse, develop a strong sense of narcissistic entitlement. This can lead to our becoming judgmental of people and things that do not fall into the realm of our likes or preferences. The antidote? We need regular reminders, no matter how small, that:

- Our preferences are not the end-all, be-all.
- We are not the center of the universe.
- Our digital experiences are shaped by our own habits rather than by objective, non-personally tailored digital histories.
- Newness, novelty, and diversity are important agents of growth.

A second impact of the algorithms created by our online histories is the potential they have for keeping us stuck in certain ruts of thought, action, or interest. For example, we might spend some time concerned about a health issue and search online for answers. Perhaps we go through a conspiracy theory phase or overindulge in an eccentric curiosity. Maybe we spend inordinate amounts of time calculating our caloric intake or tracking the latest diet trends. We "like" something, and we get more of it. Ultimately, our initial search is rewarded by a never-ending stream of more, more, more. We might appreciate this at first, but over time the incessant pitching and pushing can be destructive. Neuromarketers and app developers as well as content creators do everything in their power to make it nearly impossible to walk away or to resist their pitch, even if we no longer desire the product or information. Try to push past the feedback loop? Not so easy!

This is especially vital to bear in mind when considering the digital engagement of children and adolescents. Kids are curious by nature and grow by seeking information and interacting with it. Given their near-constant access to online spaces, young people begin creating algorithms early on. Once they've searched a topic, it's part of their history and informs what the internet presents to them. But searches about drugs, eating disorders, self-injury, and more might be best not revisited, especially when it's so hard to disengage them as habits as it is. When our past searches determine the ease, availability, and automatic nature of what we will be presented with online in the future, we are simply not released from old interests in the same ways we have been in the past. We subsequently may have a difficult time walking away from those curiosities that have the potential to harm us or hold us hostage.

FOMO: FEAR OF MISSING OUT

As noted in the last chapter, fear of missing out (FOMO) can impact us relationally. Likewise, FOMO has a profound reach into how we develop and relate to our own selves. The truth is, FOMO is a huge source of anxiety for a bulk of us who engage in digital spaces, actually impacting the way we spend our solitary time. The sense that, no matter what, we are missing some kind of connection, information, or experience means that when we do step away from our devices there is an ever-present awareness that we are "out of the loop." This makes it incredibly difficult to feel at peace and grounded with our own self.

People who play MMORPGs (massively multiplayer online role-playing games) experience this when they are away from their games—so much so that they continue to be preoccupied with their clan's (or the online world's) activities while they are offline; some serious gamers even set alarms to wake them to check in on the game when they would otherwise be sleeping. Similarly, the socially networked scan their contacts' walls and catalogue the many activities and events to which they were not privy, often feeling deep sadness or anger for having been excluded. Students, working to juggle the myriad demands placed on them by academic and extracurricular activities, toggle between digital platforms of all kinds, making sure they have left no stone unturned regarding potential for engagement. While we have always had the opportunity to feel fearful about not being included, about missing out, or about being taken advantage of, the ways in which we can currently confirm our fears and extend ourselves into multiple settings at the same time are unprecedented.

Consider Scott. Scott is the communications director for an arts organization in a large city. He is a colorful person with a diverse group of friends. He is well respected among his peers and juggles an interest in theater and performance in addition to his work. Somewhat accidentally, Scott found that he greatly enjoyed engaging with a simulated community online, where he quickly became promoted to positions of leadership. As his prowess in the space increased, so did his influence. He accrued connections and power and used his strategic mind to lead his online community to amass wealth and power globally. As this occurred, he began to spend increasing amounts of his offline life thinking about the online community. Over time, Scott began to bring his personal laptop to work so he could sign into the game during his lunch break. Soon this bled into his work, as he became unable to live with his increased anxiety about what was happening in the community while he wasn't online. Scott eventually began sleeping less and less, investing more and more time and money into the online community, all the while attempting to maintain his embodied life as best he could. In time, the online community imploded and Scott's vocational negligence caught up with him. He found himself dealing with feelings of deep sadness over the loss of both his online community and his embodied work reputation.

While this story may seem extreme, it highlights the many ways in which we alter our days so as not to miss out on opportunities for online connection, for attaining feelings of being a part of something bigger than ourselves, for entertainment and escape, and for information. Our fear that we are missing something—and might pay a price for that—drives much of our constant checking in. When we do this and don't find the responses, feedback, or further invitations we had hoped for, our anxiety and sadness drive us to fish even more. Over time, a self-feeding cycle of noticing that which we do not have, but deeply long for, becomes ingrained.

Psychologically speaking, we tend to gravitate toward information that supports what we already believe. If we believe we are well liked and meaningfully connected, we tend to scan the environment for evidence that supports this belief and hold on to it. In this pursuit, we often don't even recognize the data that conflicts with our basic belief. Similarly, for the bulk of individuals with whom I speak, a constant and pervasive fear of missing out is very real. Given this, I find that many of these folks scan their digital environments, finding evidence of the ways they are missing out or missing the mark. In doing so, they overlook and disregard evidence to the contrary. This has a deep impact on the self, passively validating the cultural belief that many activities, large social groups, and conformity to the "culture of more" is always better than their alternatives. With information gathering, we tell ourselves that being informed is more important than being calm and at peace—and we wear ourselves out staying on top of every news source and data point we can. In other words, we rely on an external locus of control rather than an internal locus of control. (I will delve more deeply into the importance of cultivating an internal locus of control in chapter 8.)

Another impact of the fear of missing out is that it drives us to overextend ourselves in ways that take us out of the present moment we embody. We all have heard about the overscheduled nature of childhood in the United States. I believe this extends to the majority of adults as well. Along with the internet comes the ability to find a plethora of ways to engage people, places, and ideas. This is truly a beautiful thing. It can also, however, push us to engage beyond what might be balanced, given that the development of a grounded sense of self requires some time for self-reflection and soothing. Until we are ready to embrace rather than fear missing out, it will remain difficult for us to develop the kind of attendance to self that affords us the healthiest lives.

TAKE ACTION

Ideas for Fighting FOMO

TAKE BREAKS FROM THE NEWS AND SOCIAL MEDIA AND COMMIT TO NOT "CATCHING UP."

Many people try to catch up on what they've missed when they take breaks from the news or social media. Instead, after taking a media break, try to pick up in the present moment, not scrolling back over what you might have missed. If you fear you might miss important details from friends or information from the news, let your friends or your inner circle know in advance that you are taking a media fast so they can inform you in other ways or at least understand if you miss responding to something big.

CONSCIOUSLY WORK THROUGH FEELINGS OF BEING LEFT OUT.

When we feel confident in our ability to handle our feelings, we are more resilient in the face of perceived or real rejection. Figure out how you can best comfort your feelings of rejection. Do you need to cry? Yell? Go for a long walk? Can you work through the discernment process of how big a deal it actually is that you weren't included in a particular event? If missing something feels significant, see how you feel after a day with some real processing and then decide if you should talk with the host. If not, can you comfort yourself in meaningful ways to bring your reaction to a fitting "size"? This process is especially important for children. Rather than trying to console your children by saying, "Don't be upset. You'll be invited next time," help them to process their feelings in response to being left out and to learn from the experience. This cannot be overemphasized.

AFFIRM THE PLACES AND PEOPLE TO WHOM YOU ARE MEANINGFULLY
ATTACHED AND INVESTED IN.

When we take the time and energy to identify the places where our
involvement and investment truly matter to us and the world, we
may become more able to keep the places where we aren't involved
"right-sized," therefore not needing to grieve for those things we
miss out on that may not matter as much as we imagine.

FALSIFIED AND CONSTRUCTED SELVES

Much has been written about how our presentation of ourselves in
digital spaces mimics or differs from the way we present ourselves in our
embodied lives. Whether we interpret this as a positive or negative reality, we can likely all agree that digital spaces allow for a shaping of the
self that our embodied spaces will never afford. In ways large and small,
conscious and unconscious, when we present ourselves online, we make
choices about what we include and what we hide. While this is true in
our embodied lives as well, the opportunities for *control of the image* we
present are much greater online than they are in embodied experiences.

No longer tethered to cameras with film that needs to be rationed
(only twenty-four shots to a roll!), processed, and developed, we take
hundreds of photos until we find the one that most captures the look
we are going for, and then we post it. The preponderance of filters and
self-improvement apps tells the story of how much many of us would
like to present our best (and often altered) selves online. We add to this a
carefully curated cadre of information about us and, voila, we have constructed a self. For many of us, the authenticity of the information is not
what dictates inclusion of a photo or profile detail; instead we are driven
in our decisions by what will garner notice, attention, and the much-
sought-after "follow."

When the image we create of who we are is closely correlated to who
we are in embodied spaces, the likes, follows, and interaction we receive
in these spaces feel congruent. But when there is significant discrepancy
between these two selves—by intention or lack thereof—a sense of internal conflict occurs: we know we haven't been fully honest about the self
we've projected, yet that entity is engaging with others and receiving
real feedback and connection. When this happens, we are faced with the
difficulty of determining whether our authentic selves, which have been
misrepresented in the digital sphere, are as desirable and acceptable as

those we have created. This can create a serious conflict for us about how to be and what we can rely on as "real" in our internal and relational lives.

Recently, my cousin, a fourth-grade teacher in a suburban school, showed me the app for a community football team on which some of his students play. In addition to being a collection place for information regarding game times and locations, the app boasts "player cards" for each fourth grader on the team. When accessed, these cards display a highly stylized photo, taken from the ground up, with the well- suited player standing in a power pose in front of a cloud of smoke. This perspective makes the boy look tall and menacing, which is also reflected in each boy's facial expression.

These hypermasculine photos are the perfect example of the way in which we construct images of ourselves in digital spaces. There is no way that any fourth-grade boy is tough and strong *at all times*. When, however, such a child gets the kind of attention he will likely garner for this very image, he subtly internalizes the message that this version of himself is the one that will be most rewarded by attention. This is a hard message to undo, especially for a child. When we teach this lesson to ourselves by fostering online images that differ from our actual embodied, raw selves, it's easy to slip into a dependency on others for affirmation and approval. This is the opposite of developing a whole and hardy self.

Another complexity is that we commonly construct slightly different versions of ourselves across digital platforms. I cannot count the number of stories I have heard about people who have maintained squeaky clean online presences in those spaces they thought to be public, only to find themselves denied a job based on something a potential employer found in a less-prominent place online. A random, edgy (and often completely congruent) tweet, a long-forgotten Myspace account, or more can end up undoing even the most carefully constructed profile. With the complexity involved in maintaining so many places and spaces, it is not surprising this occurs as often as it does. Again, this dynamic is at play in our embodied lives as well; it's just that its digital incarnation is much more widely available, easily shared, and less in our control. Having a healthy sense of self is challenging enough as it is. Trying to traverse a world where we have created multiple versions of this self is excruciating. It's important to realize this for ourselves and others, and to graciously invite both to be authentic and congruent.

A TENDENCY TOWARD A FIXED MIND-SET

Carol Dweck of Stanford University has done extensive research on what she has coined growth and fixed mind-sets. In a *fixed mind-set*, people believe there is a certain hurdle to cross before attaining any given status. This way of thinking says, "To be smart, you must receive A's," or "To be considered successful, you must come in first place." It assumes or prescribes a set marker and predicates status on achieving or surpassing that marker. Individuals who adopt fixed mind-set models tend to peak early in life in terms of success and can become risk averse, not wanting to threaten the successes they have achieved by shooting for higher markers and possibly failing. These individuals live to standards largely outside themselves and revere accomplishments more highly than effort or improvement.

Most American academic and vocational models work on a fixed mind-set model. If we do certain things, we attain higher status. If we hit certain targets, we are rewarded with certain privileges and distinctions. To be sure, much of the online world thrives on this kind of mind-set as well. We attain a certain number of followers, and we feel we have arrived. We strive to reach higher levels and status in our games by meeting set standards. We see certain images and standards of beauty or gender or strength or popularity, and we work to achieve membership among these ranks. In short, fixed mind-set models leave us at the mercy of finish lines we must cross.

A *growth mind-set,* on the other hand, is based on the idea that appropriate risk-taking, even failure, can lead to success. A growth mind-set says, "Try things that may be a stretch for you. As long as you put in real effort, you will learn, even if you don't accomplish what you imagine is 'success.' Even failure can teach!" Growth mind-set models encourage effort and dedication to a task; they foster grit and resilience and lead to higher levels of satisfaction. In addition, they value appropriate risk-taking, see failures as opportunities to learn, and reward effort over mastery. A person with a growth mind-set says, "I can handle the consequences of failure, so I am willing to take a risk to learn, advance, or grow." Granted, some games and other digital platforms are working to promote this kind of thinking. For example, *Minecraft* works on real-world physics principles, providing encouragement to build objects that will "work" without the high stakes of failure or demotion when it takes multiple attempts to get something right. Nonetheless, a great number of the online spaces we inhabit do not provide the kind of grace-giving encouragement that a growth mind-set ideally offers.

This is especially true in the social networking arena. Because of the way we consume information on these platforms, we frequently take in the data we find there and convert it into fixed targets against which we compare ourselves. This happens because the preponderance of content we find in these spaces comes to us bereft of meaningful context. We read a post and ignore the reality of the writing and rewriting that made the post "just so." We see the many social connections and outings of those in our networks without becoming conscious of the many times those same people are alone or missing out just like we are; they just aren't posting about those times. We see an image and do not do the work of considering the editing that went into it: lighting and prepping, scouting of site and application of filter, posturing, makeup, and possibly even Photoshopping. In fact, we are taking in images with immense production values (not always monetary) and internalize them as though they are the new norm. Once we've established this new standard, we compare ourselves and our lives with them. In this way, we subtly reinforce fixed mindset thinking, telling ourselves that we have either "arrived" or we have not. This keeps us not only pushing toward standards outside ourselves but also locks us into a cycle of thinking that the ideal is the norm and provides a constantly moving target we can never reliably hit. So many negative emotions, including envy, can arise from this unhealthy cycle. Consider the perfected personas of their families that many parents post. How can those of us struggling to get through the crisis of the day with our children not feel both envy and self-doubt about our own abilities as a mom or dad when we see online that other families (according to their posts and photos) have the perfect children, the perfect house, the perfect jobs, and more?

TAKE ACTION

Ideas for Enhancing a Growth Mind-set

Try new things without overemphasis on mastery.

It is important, at least some of the time, to try things we are not already good at. Don't worry too much about what you try, just do something new! Listen to a new genre of music, try bowling, take a new kind of yoga class, skip instead of walk, walk or run in a new neighborhood, navigate with a paper map, take an art class, try origami. Honor and affirm your effort, not the outcome.

Master a useless skill that takes time to learn.

Commit to building a new skill that will require time and effort but that doesn't get you anything in return. For example, try learning to sign your name with your nondominant hand, or take up juggling or kendama or hula-hooping. Once you master that skill, try something more challenging. Notice how it feels to fail enough times that you eventually succeed.

INSTEAD OF JOURNALING, TRY A BRAIN DUMP (STREAM OF CONSCIOUSNESS WRITING).

Write down your thoughts on a piece of paper for five to ten minutes straight. Don't try to construct sentences or build ideas. Simply write whatever comes to mind. When you are done, rip up the piece of paper or burn it. The process, not the outcome, is the goal.

BOREDOM INTOLERANCE

Feeling at peace and grounded in one's own being is a hallmark of a healthy self. One of the most beautiful rewards of living from this grounded center is the ability to tolerate boredom, sometimes even to invite it. Boredom offers space to explore and forces us to encounter our very selves. When we are free from stimulation that distracts us, we are brought to the end of our hiding and into new spaces of exploration. When we have nothing to look at, listen to, or engage with, we are given the opportunity to see what we ourselves are made of. For some, these times come as ready breaks and afford opportunities for rest and refueling. For others, open spaces usher in creative thinking and active exploration. For many of us, however, these times catch us off guard and unprepared, so we scramble to avoid the discomfort of feeling purposeless and without direction or distraction. If we have not yet come to value free time, if we do not have the skills to use it effectively, or if we indulge the impulse to fill all available space, we often fritter away open spaces and miss out on opportunities to encounter ourselves in new ways. Ultimately, we lose out on all the benefits of just being.

Unfortunately, boredom isn't rewarded in our culture; hard work and productivity are. I find that many of us have internalized the value of hard work and productivity in passive ways that may not be in keeping with our culture's original drive toward production. We feel a greater sense of productivity when we surpass another level of our current online game than when we allow ourselves to be still and quiet for a time. We answer just one more email, we troll just one more Pinterest board, or we watch just one more TV series episode—all with the unconscious sense that because these actions require us to *do* something, they accomplish the unconscious goal of keeping us productive.

In fact, allowing space for boredom is frequently experienced as a risk. We have linked boredom with distractibility and claim that we must stay on task and constantly plugged in to be successful. This is reinforced by work and educational settings that run on twenty-four-hour-availability norms, subtly sending the cue that, if we are bored, we are certainly falling down on the job. As our unaccounted-for time is often what we use to engage in the least beneficial spaces on the internet, we also *fear* boredom in some cases, as it leads to our feeling shame about our "bad behavior." Linking our lack of comfort and familiarity with boredom to the threat of failure, harm, or shame easily explains why we shy away even from meaningful discussions about it!

The benefits of finding and protecting time to be centered and bored—and letting it teach us—could be countless. Boredom tolerance has the power to reduce our anxiety regarding missing out as well as the power to develop our capacity to value ourselves for who we are rather than for simply what we do. This emphasis on *being* over *doing* allows us to live from a place of grounded confidence, knowing that we can handle a lack of stimulation without anxiety. Also, research finds strong correlations between boredom tolerance and creativity. The evidence is clear: Boredom benefits us. But to tolerate it, we need a store of resilience.

TAKE ACTION

Ideas for Increasing Your Comfort and Experience with Boredom

TURN OFF NOTIFICATIONS.

Sometimes not knowing you have a text or email waiting will allow you to settle more into "you" in between moments than if you look at your phone and see a screen full of notifications.

SET A PASSWORD.

Even the extra step of having to enter passwords makes us less likely to be overly attentive to our phones.

LEAVE YOUR PHONE IN THE TRUNK OF YOUR CAR.

This works while you're driving, out to dinner or an event, and when you're at home. When your phone isn't as handy, you are likely to force yourself to go without it and embrace your embodied present—in all its stickiness, messiness, discomfort, or joy.

HOST A BOREDOM PARTY.

This is a high-level challenge, requiring you to take big risks and tolerate whatever consequences ensue. Invite friends over for a boredom party. No one can plan anything specific ahead of time or coordinate who is bringing what. Prepare your guests that the time will begin with everyone simply "hanging out" with no particular purpose. Over time, see what is created out of a gathered group and some free space and time.

UNDERDEVELOPED RESILIENCE

"Resilience" refers to the ability to handle difficulties and hardships without facing psychological symptoms. This trait has gotten a great deal of press across many disciplines, with experts suggesting its serious decline in children and young adults. Tech use is listed as a potential mitigating factor. In *Boys Adrift*, for example, noted family physician, psychologist, and parenting expert Leonard Sax contends that video games do not engender the sense of resilience or the patience and drive the real world requires. And it isn't just video games that put resilience at risk. Social media, instant access to information, and any number of other platforms and devices may rob us of exercising and cultivating this important trait.

Many relate the decline in resiliency to helicopter parents hovering to protect their children from real, perceived, and potential difficulties or failures. While this may be true, it may be overemphasizing the role of parents' rescuing behavior and underemphasizing the way in which our devices rescue us from our current realities. When we face hardship, we can escape into all manner of digital content rather than face our challenges head on. When we find ourselves uncomfortable in a conversation or interaction, we can conveniently remove ourselves from it via technology. When we can't solve a puzzle, answer a question, find a solution, we can turn to Google and move on without ever having to wrestle or dive deep into our own problem-solving capabilities. The reality is that when we feel uncertain about what we can and cannot manage and achieve on our own, we live from a kind of brittle tentativeness that impacts the way we engage in our embodied lives. This is diametrically opposed to the developmental process of embodying a sense of resilience.

When we have developed a sense of resilience, we feel confident to handle the consequences of our actions and even our failures. We know we have what it takes to comfort ourselves if we fail, and to get back up if we fall. Unfortunately, our digital engagement has the potential to rob us of the kinds of experiences we need to develop even small amounts of resilience. When we can order and pay for our meal online then pick it up without making eye contact or speaking to anyone—and then replicate this experience across our daily life—we begin to lose our sense that we can handle the awkwardness of being an embodied self out and about in the world. For example, sometimes it benefits us to place a food order human to human, with eye contact and bumbling words. If we pay with cash and end up dropping our change and needing to gather it up, all the better. Sometimes it benefits us to get lost and figure out our own way, even better if we strike up a conversation and consult a friendly bystander (rather than consulting our phone). Living through awkward life experiences is the currency of resilience development! When we avoid these situations, what might have been considered small, awkward moments over time become looming, potentially risky social experiments, simply because we lack the resilience that everyday embodied encounters build through practice over time.

TAKE ACTION

Ideas for Building Resilience

DO AN INVENTORY OF THE FEELINGS YOU ARE COMFORTABLE WITH AND THOSE THAT MAKE YOU UNCOMFORTABLE.

By identifying those feelings you can comfortably tolerate and those you avoid, you can begin to work specifically and deliberately to build your competence and confidence. If you know, for instance, that anger is an emotion you actively and passively avoid, spend some time creating tools for managing anger when it arises. You might make a list of such ideas that reads: "When I face anger, instead of avoiding it, I will build resilience by facing it. To do that I will . . . (e.g., express my anger by taking a fast walk/jog around the block, getting into a safe place where I can yell or cry, doing a quick "brain dump" then tearing up the paper, or throwing ice cubes against the back fence or onto the cement)." When we have a plan for how to deal with complex or difficult emotions, we are less likely to avoid them and our resilience builds.

DO A DAILY EXAMEN.

An "examen" is a practice found in many religious traditions. Some people call it a "Rose, Bud, Thorn" exercise; others call it the

"Crappy/Happy" exercise. I suggest keeping a small notebook or simple piece of paper next to your bed. Each night before going to sleep, record what gave you life during that day ("happy") and what took life away ("crappy"). In the Rose, Bud, Thorn version, you would name an experience that was beautiful (the rose), an experience that was painful (the thorn), and an experience that you are looking forward to the next day (the bud). Over the course of two weeks, you will begin to see what experiences zap your energy or turn your day. Begin to strategize about how to handle these situations differently. This is directly making a plan to build your resilience.

DYSREGULATION CAUSED BY SUBSTITUTING STIMULATION FOR SOOTHING

When we look to devices to entertain and soothe us and rely on connections in the digital domain to inform our feelings, thoughts, and attitudes about ourselves, we are at risk of being thrown off at any moment. A friend's celebratory post on social media prompts feelings of comparative inadequacy. A clan's defeat due to a personal misstep or miscalculation initiates feelings of anger. The photo we posted with certainty of a huge response goes unnoticed, and we feel deep disappointment. As this happens, our days turn on occurrences outside ourselves. This has always been true. Circumstances, experiences, and people in our environments have always been able to derail us. Never before, however, have we had such constant access to the world outside ourselves.

Life hands us many experiences throughout our day. By virtue of our very humanity, our own temperament and our personality provide us with all manner of *internal* experiences throughout the day as well. Sometimes the external situations create internal responses and vice versa. The ways in which we influence our experiences and our experiences influence us are complex. These interplays can result in an everyday steadiness and confidence if they are in concert with our thoughts, emotions, and physiological states. If, however, there is a mismatch between the situations we are in and our very thoughts, feelings, behaviors, or capabilities, we may find ourselves feeling somewhat derailed and disintegrated. This internal sense that things are a bit "off" can be referred to as a "dysregulated" internal state.

Each of us can turn inward or outward for assistance in steadying ourselves when we find ourselves dysregulated. Optimally, we find healthy ways of resolving the mismatches in our contexts and internal worlds. We stop what we are doing when we find ourselves in this state, assess what is needed, dig deep to find some soothing or confidence, seek out reliable and trustworthy supports, and work to regain regulation between all the disparate parts of ourselves and our experiences.

The trouble is, instead of relying on a healthy process of internal self-regulation, we now rely more consistently on our devices to regulate us. When we experience a difficulty in life, we turn to our phone for answers, distraction, or access to others who will do the work of soothing (or distraction) for us. The vicious cycle created by living fluidly between digital and embodied spaces is real. Defaulting to our devices can begin to feel like the easy way out when we are the least bit restless or challenged by life. In addition, by constantly checking in with our devices, we miss out on developing a rhythm that includes checking in with ourselves or the moment at hand. This subtly reinforces the idea that our devices offer us a greater sense of grounding and are worth coming back to more than our actual selves are. Over time, we develop an external locus of control and find ourselves with underdeveloped internal regulation skills. Chapter 8, "Cultivating an Internal Locus of Control," will explore this concept in depth.

Given the reality that the internet can transport us anywhere at any time, coupled with our well-honed drive to multitask, we have become highly distractible and distracted people. When we want to avoid a task, we now have unlimited places to escape to. When we aren't getting the attention, validation, or stimulation we need from the people around us, we can turn to digital spaces to garner it. We read reviews ad nauseum to avoid making decisions about how to purchase or invest, and we fritter away idle time checking the multiple online spaces we inhabit. For most of us today, rapid movement between both worlds happens seamlessly and without apparent effort. But to have a healthy internal locus of control, operate from a regulated state, and enhance our character, we need to develop the ability to be in one place at a time.

What happens when we offload our regulation to internet-enabled devices is, basically, a bait and switch. We need soothing, but we substitute stimulation. We need to get calm and centered; instead we gather more data, input, and dazzling digital experiences. This leaves us dependent on stimulation to distract us and make us think we are actually being soothed. On the contrary, being soothed results in calming and working through the feelings related to dysregulation. When we substitute simple distraction and stimulation for this developed ability, we end up amplifying the dysregulation we are already experiencing and rob ourselves of practice in the important work of bringing ourselves back to a regulated state.

Complicating things, without consciously realizing what is happening, many individuals develop patterns related to their engagement with digital platforms that actually result in dysregulation in and of themselves. Three such patterns come to mind as especially relevant here:

- Dysregulation related to using social media networks.
- Dysregulation related to engaging with violent forms of media.
- Dysregulation related to consumption of idealized images of bodies and objects.

When interacting with online social networks, many individuals experience a variation of dysregulation without a conscious awareness of the dynamic. By interacting with socially networked platforms regularly, they develop strong habits that make the online network an increasingly embedded part of their daily routine. While these individuals may enjoy parts of their online social experience, those at risk of being negatively impacted will often become excessive in their use—often disregarding feelings of inferiority, envy, or upset that result from engaging the networks. This phenomenon has been verified by research. According to studies, heavy users of social media are 2.7 times more likely to experience depression than less-frequent users, and those engaging social media for two hours a day double their feelings of social isolation. This demonstrates how the dysregulated state of needing affirmation and recognition outside one's self or of comparing oneself excessively to others can easily become a source for greater dysregulation, as one works harder and harder to find *externally* what optimally one needs to master *internally*.

Emotional, mental, and physiological dysregulation is common particularly among those engaging with violent forms of media. Most research on this topic relates to video game play. Some studies have found strong correlations between such play and elevations in the release of neurotransmitters and hormones that cause irritability well past the time of play; other research has shown that the immersive elements of video game play can cause physiological fight or flight responses for the player. This means that a person in a nonthreatening physical space may reside there in a deeply dysregulated internal state as a result of the consumption of violent media. Amped up and unaware of how to soothe the self, this individual may want to loop back to the immersive land of the game to avoid his or her own internal dysregulation or encounter the blandness of the embodied physical space.

Like those who play violent video games, another group at high risk of dysregulation caused by overconsumption of digital media are those who experience any kind of ambivalence or discontent with their physical appearance or capabilities. And, let's be real, who doesn't fit into this category? Never-ending streams of Photoshopped and otherwise altered images of bodies can result in dysregulation that is painfully palpable.

If you leave your house and breathe, you are flooded with these images every day. Seeing that one can never live up to the image ideal over and over and over each day is wearing. If you play video games or interact with social media or video content, the images more than flood you; they drown you.

The good news is that we can learn to recognize dysregulation for what it is, and we can gain tools to resolve it. We can learn the ways in which we get tripped up. We can discern what we need to genuinely soothe ourselves rather than simply distracting ourselves or stimulating ourselves in the hopes that the feelings of dysregulation will go away. Self-soothing is actually a superpower of sorts. When we master self-soothing—even if we must borrow from another's care for us as we master and integrate our ability to soothe ourselves—we have the ability to handle nearly any stressor thrown our way, to handle boredom, and to become more resilient. The pauses in our daily lives that are created by the ability to self-soothe also offer us opportunities to develop an internal locus of control.

TAKE ACTION

Suggestions for Learning to Self-Soothe

LEARN TO PRACTICE MINDFUL BREATHING.

If you can, add stretching. Proper breathing alone can go a long way in helping us to regulate ourselves. (I'll talk more about breathing at the end of chapter 11.)

FIND "ESCAPE ROUTES."

We often let ourselves become overstimulated because we don't know how to get ourselves out of situations. Determine places you can escape to where you can catch your breath and tap into your grounded self. The bathroom is an obvious place in most situations. Look for other places to do the same. Back stairwells in office buildings and your car are other places to which you can return for quiet and calm. Over time, practice excusing yourself in order to take a pause. Use phrases such as "I need to take a minute to gather my thoughts. I'll be right back"; "I am reacting really strongly to what just occurred. I need to step outside for a breath of fresh air"; or "I am surprised by how amped up I feel. I'm going to run up and down the stairs to exert some energy and will be right back." If you are confident about getting what you need, people are unlikely to bat an eyelash at your actions. In fact, they'll likely want to emulate your behavior.

MAKE A "SELF-SOOTHING" LIST AND REFER TO IT.

Post it for others if you can. Identify at least twenty actions you can take that are soothing to you when you are dysregulated. This list should be varied and include actions that require planning and some that can be done spontaneously. Some should be free, and some should cost money. Write down some that can be done at any time and some that are specific to time and place. Keep this list in a handy place. If you live with others, offer it to them so they can understand how you best self-soothe and can encourage you to do so. Ask your friends and partners if they might like to create and share their lists. It's easy to fall prey to the assumption that self-soothing should look a certain way, but not everyone loves massages or manicures. We each must identify and engage our own unique preferences.

GO TO A CORNER OR APPLY SOME GENTLE WEIGHT.

When we are very overstimulated and dysregulated, we may need to soothe ourselves from the outside in. In this situation, it can help to tightly wrap a blanket around your body or to sit with your body on the floor, leaning your back into both sides of a corner. Heavy or weighted blankets or large versions of the seed- and grain-filled tubes that can be heated and placed on sore muscles are also helpful in communicating to the body that there is space for soothing and stillness.

A COMPLEX ISSUE

Our modern world offers a nearly constant flow of information and entertainment. Opportunities for development of an internal locus of control, self-soothing and stimulating skills, and moderation do not necessarily naturally present themselves anymore. Because of the infrequency of these beneficial opportunities, we are often uncomfortable and sometimes even anxious when they arise. In an effort to calm our anxiety in the face of unknown quiet or opportunities to rely on an internal locus of control, we turn to our devices to occupy us. The pattern goes like this:

Need for regulation → leads to discomfort → leads to need for relying on internal processing and tapping into our internal locus of control → leads to feelings of inadequacy and the scary unknown → leads to anxiety → leads to distracting oneself with technology.

This pattern, when repeated over and over, creates a habituated response that at its root says distraction and stimulation are better than leaning into the unknown or less-developed internal world. In one well-known study, participants were placed in a room for fifteen minutes to be alone with their thoughts. Some 67 percent of the men and 25 percent of the women chose to administer electric shocks to themselves rather than simply sit quietly. These same participants had previously claimed to be willing to pay money to avoid electrical shock during research. Such research simply highlights what many of us have observed that being alone with one's self, thoughts, and feelings can be a difficult task. When alternatives are at the ready that will help us avoid the experience, we, too, often jump at those alternatives, leaving us with underdeveloped abilities in soothing and entertaining ourselves.

When a person demonstrates addictive, attention-seeking, risk-taking, or antisocial temperament traits, technology poses an even greater risk to the development of the self. Individuals with strong propensities to function as loners and those who suffer from high levels of anxiety face similar risks. These identity markers might set the stage for higher addictive tendencies with technology in general and devices in particular. These are of course generalizations, yet they are important to consider when addressing the development of the self. Anything used in a compensatory way can end up interrupting the stability of the self by becoming a point of dependence rather than a source of thriving.

This is very different from the way some assistive technologies end up escorting certain populations into identity formation and life satisfaction. Technologies designed to assist special populations with communication or to offer opportunities for social practice and increased human connection are profound offerings. Lives have been changed by these powerful tools. Many of the individuals who use these see real and lasting improvements in their embodied life. The challenge is in discerning to what level the individuals can be taught and encouraged to take what they learn in these digital spaces and apply it in the embodied world. If those supporting these individuals can help create these kinds of "transfer" opportunities, the person is doubly blessed.

The preponderance of all manner of apps and platforms for special populations can make it difficult to know what technologies to engage and when to engage them. A good rule of thumb when evaluating what and when to use these is to research the gold standards in the particular area you are seeking to enhance. With new apps and tools being released every day, it can be easy to find the newest, shiniest app with the best-sounding claims. Instead, look for a well-tested and evaluated platform or app that has reliable developers who are keeping the platform or app up to date, specifically relevant, and safe for users. A consultation with an educational psychologist or technologist who understands the particular skills that need to be addressed as well as the particular learning or life needs of the individual can also go a long way toward finding the best apps and tools for each situation. Finding even tiny opportunities within the natural course of life to encourage transfer of skills developed within these technology-driven spaces can be profound.

ADDITIONAL IDEAS FOR NURTURING A MORE GROUNDED SENSE OF SELF

CONSIDER A CONSISTENT SOCIAL MEDIA FAST.

Media fasts of all kinds are encouraged in lots of places. While a one-time period of going without phones or other devices can be very illumina-tive, I find that the impact wears off relatively quickly. Instead, consider choosing a period of time during each day or once a week that you can go without your device with relatively few consequences. Set an alarm to remind you to do this until you get into the habit. Create a ritual of sorts that you engage when you turn off your phone, as opposed to simply silencing it. Be mindful that you are taking a step toward being more calm and less distracted. Start and end your time by taking a few deep breaths and a few gentle stretches. In the early times of doing this, you may want to keep a notepad handy to take notes or make lists of things that come to mind.

BUILD A VOCABULARY FILLED WITH NONEVALUATIVE, NONCOMPARATIVE LANGUAGE AND EMPATHIC, ENCOURAGING, AND LIFE-AFFIRMING SENTIMENTS.

Intentionally seek out new words that are positive and life affirming. Find mottos and mantras that are infused with truth and inspiration. When you find yourself looking to others to shape how to feel about yourself, work to resist this and try to identify something you appreciate about yourself. If you find this usually involves evaluative language, try again. Try to find noncomparative ways of considering yourself and others.

TRY THE "HALT" SCAN.

Borrowed from twelve-step programs, the HALT scan involves stopping either throughout the day or when you feel particularly dysregulated and asking yourself if you are hungry, angry, lonely, or tired. These four states of being leave us particularly prone to distracting ourselves or to using things other than what we really long for to satiate us. Once we've identified any areas that need to be addressed, we can choose a best path rather than simply acting unconsciously. If you are lonely, it is easy to grab your device, without even thinking, and shoot off a text or two. Sometimes this may be exactly what you need, but at other times you may need a more in-depth encounter with someone to truly meet your needs. The same is true of the other states of being. The more intentional we are about identifying our current conditions and addressing them for what they truly are, the more we live from an integrated core self. This also is a great tool to teach children; it will serve them well throughout their life.

Part II

Moving with Intention toward a Grounded Self

7

The Fertile Ground of Idle Time

I spend a lot of time in my car and on the move. Periodically, I'm guilty of leaving my car running while I wait in it. I do this not so much on purpose but instead because I just don't think to turn it off. In these times, the car sits idling with the engine running, but the car isn't moving. My first car, a pea green 1970 Ford Custom 500, idled more actively than any car I have ever driven. As it sat, its motor would rev higher and higher until I would, with the car in "park," punch the gas pedal, inspiring the engine to slow down. In that car, I could tell things were happening even when I wasn't moving. The same is true for people. While the body is still or otherwise disengaged, much continues to go on inside.

Most dictionary definitions of "idle" reference words such as "unemployed" and "inactive." Others dip into descriptors that feel value laden, such as "lazy," "loafing," "unimportant," and "worthless." When applied to the human condition in the developed West—where the values of productivity, accessibility, speed of response, and breadth of impact are highly valued—these latter definitions could suggest that being idle is a negative condition, a waste of time. This makes me deeply sad.

As today's basic high tech products boast about the way in which they might make life easier and free up additional minutes in our days, the *potential* for encountering idle time might increase. Yet the same resources that tout improving productivity (such as mobile devices, high-speed communications technology, and networked communities) also provide easy, seamless access to time-"wasting" opportunities such as web browsing, shopping, and mindless entertainment. Ironically, then, the "conveniences" that create the idle time often automatically claim the newfound free time before a person realizes the new time exists.

Before our world was hyperconnected, there were unavoidable and built-in opportunities for our bodies and brains to experience times of idleness. While traveling from one place to another (by foot, horse-drawn wagon, bus, car, or subway), we had to rely upon that which was *in* us to engage us. Our thoughts, feelings, and physical sensations plus whatever we could carry (a book, a notepad and pen, or a newspaper) were our only aids in passing time. The same was true for most of our daily lives. Moments not taken by work or tasks provided opportunities for inactivity, for idling. This created rich windows of time for reflection, critical thinking, daydreaming, and rest. Confronted with moments of seeming nothingness as a regular part of normal days, people were gifted with consistent opportunities to be still, to become comfortable in their own skin, and to be good company for themselves. As naturally occurring pockets of inactivity have disappeared from our daily lives, so have the opportunities for serendipitous stillness and the creative idling space these states of being offer.

Idling actually has immense potential to command our attention. When we are in constant intellectual, emotional, or physical motion, we lack the spaciousness needed to come to understand and make sense of the full richness of our humanity. We are all familiar with the experience of feeling hungry or tired and not paying attention. Our stomachs growl or we yawn, yet we mindlessly push forward. We might drink coffee or eat something out of the vending machine, whatever is needed to keep moving through our very full day instead of taking the hunger pains or feelings of fatigue under real and rationed consideration. Our cultural norms reinforce this compensatory pattern by rewarding constant productivity, action, and advancement. As such, we are most commonly validated for having our attention focused *outside* ourself. Not only are we rewarded for being available to our employers, educators, and social connections twenty-four hours a day, but we are also privy to a never-ending stream of entertainment, education, and information that feels as though it builds, soothes, or stimulates us. Little reason (let alone, demand) exists anymore for using our idle time to turn our attention *inward*.

In many ways, we have grown unaccustomed to intentionally choosing what we pay attention to. Instead, we most often find ourselves attending to whatever happens to capture our attention or take hold of it. We habitually end up noticing that which is most present, urgent, unconscious, and, basically, shiny. Because we are constantly presented with new stimuli originated from complex algorithms that are tailored, very specifically, to our interests and behaviors, the pull on our attention is uniquely compelling and hard to resist. As a result, we invest increasing amounts of our time in interacting with these spaces, all the while consciously or unconsciously developing an *external* persona.

We make gains as well as pay prices for this massive investment of time in our externally developed, stimulated, and soothed selves. It is widely believed that if we practice an action we will increase our skill. Meeting free time with a practiced turn toward demands or opportunities outside ourselves thus increases our skill at attending to our external selves—and leaves little time for developing a grounded, reliable internal gut. This results in a strong knowledge of what is expected of or available to us— and a weak knowledge of who we are at the deepest level.

WHAT STILLNESS AFFORDS

How we use our idle time is perhaps one of today's greatest unasked questions and has huge potential to enrich our world personally and communally. Taking idle time to be still, to evaluate the genesis of our feelings, say of hunger or exhaustion, and to attend to them in healthy ways is to live counterculturally. To take a break from demands in order to deliver healthy fuel (in the form of nutritious food, meaningful rest, or fulfilling creativity) takes more time and energy than meeting them in less intentional ways and may actually bring with it consequences from those who do not honor time for such endeavors. It is to live by a standard that is higher than the norm!

Matthew J. X. Malady, a columnist for *The New Yorker*, wrote about his three-day digital fast in a piece titled "The Useless Agony of Going Offline." Motivated by the work of technologist and professor Daniel Levy and by hearing about a man who fell to his death by stepping off a cliff edge while looking at his cell phone, Malady decided to power off all devices from New Year's Eve to January 3. While he reports that he got more things done and felt an expansion to the length of his days, he reflects primarily on what it felt like to go without information-seeking tools. The honesty of his reflections is profound:

> So many questions went unanswered during those seventy-two hours—so many curiosities cast aside and forgotten without being pursued. I was less harried, I suppose, but I was also far less informed, and not as advanced in my understanding of all sorts of things that interested me. I felt as though I were standing still rather than moving forward. And while standing still for a while can be pleasant, it's not without its drawbacks. Instead of feeling more relaxed, I mainly felt unfulfilled.
>
> I would like to say that I reached some time-maximization epiphany during my New Year's experiment, but I'm not sure that I used my time any "better" than I normally would have during that span. I just used it differently, and found myself frequently bored as a result. . . . At the end of the experiment, I wasn't dying to get my phone back or to access Facebook. I just wanted to get back to being better informed. My devices and the Internet, as much as they are sometimes annoying and frustrating and overflowing with knuckleheads, help me to do that. If getting outside and taking walks, or sitting in silence, or walking dogs, or talking with loved ones on the phone got me to that same place, I'd be more than happy to change things up.

I can relate to Malady's takeaways. In a world where information is available and time moves rapidly, attaining quick answers is invaluable; there are many times when fast and thorough information access makes life feel and seem better. Reviews lead us to the most prudent purchases and experiences, and googled answers inform our discussions, thoughts, activities, and interests in personally meaningful ways. Although not a negative thing in and of itself, quick access to information is also not completely benign. When we habitually turn to our devices for answers or stimulation, we send our brains and bodies the message that information and entertainment culled from external sources is of a higher value than information and stimulation that can be accessed from our own internal resources. Basically, access to a constant stream of information and entertainment can lead to a diminished value in or ability to access and live within the bounds of one's own thoughts, ideas, and experiences. It also stops curiosity and, at times, cuts creative problem solving short.

Life today offers few naturally occurring opportunities to develop the skill of utilizing idle time in the service of being still. This lack of opportunity for and practice with engaging spaciousness makes us greet it with anxiety and stress. This leads us to avoid it in both deliberate and unconscious ways. Rather than protecting and saving idle time, engaging the stillness it might offer us, we reflexively fill it up. We listen to curated soundtracks as we walk or run, watch YouTube videos while we use the toilet, binge on podcasts while we mow the lawn/do the dishes/brush our teeth. We spend our days scrolling through Facebook and Instagram feeds, pinning on Pinterest, and tweeting what we notice as we stand in line. Our embodied interactions are interrupted by texts, news updates, and all manner of digital notifications. This leaves us bereft of time to practice the art of engaging idle time for personal development. Without such practice, we lack the skills necessary for tolerating and proactively using times of idleness, leading us to avoid the unknown.

The costs don't end here. Our intolerance for stillness leaves us with a resulting deficit in self-soothing abilities. As discussed in the last chapter, we cannot be still because we can't soothe ourselves, quiet our thoughts, or regulate our emotions. Instead we stimulate ourselves, distracting ourselves or denying our need for comfort. Without the developed ability to engage idle time, we end up seeking more engagement with our devices by default. This cycle causes us to avoid the very experiences that might build our capacity to bank idle time in order to practice tolerating stillness and use it to benefit our ability to know and engage ourselves.

INWARDNESS

I often think about ways we might be inspired to stop our constant online searching and engagement and instead protect and engage even small moments for getting to know ourselves better. I wonder what kind of self-knowledge might result if we left our earbuds at home or sat on the subway without engaging a device. I imagine potential benefits gleaned from the simple act of standing in line without pulling out a phone. If we were to do these things even some of the time, I believe we would enlarge our ability to genuinely be with ourselves.

Being able to be with ourselves in all our complexity requires an ability to be still at least occasionally. When we fritter away all our moments in unconscious searching for answers and entertainment, we are left with few moments of idle time within which to develop the skill of stillness. And developing this skill is rife with profound potential! This isn't just common sense; it's also found in scientific studies that demonstrate correlation between boredom tolerance and higher levels of creativity and focus, two skills born from the ability to tolerate stillness.

Consider people who don't need to fact check in the midst of a person-to-person interaction, those who can describe a situation without needing to pull out their phone to show you photos, and those who don't need constant input to move through their days. People such as this, who can tolerate quiet or the simple lack of stimulation, find capabilities that positively impact their sense of satisfaction in life and relationships. When we feel capable of tolerating the unknown, a steadiness occurs. When we can handle sustained focus and have the ability to delay we become empowered with a sense of internal regulation.

Stillness is that nondirected present-ness housed within an alert body that allows for a rich ability to pay attention to both our internal and external realities. It can take on many forms. Stopping physical movement is one way of being still. Halting distracting thoughts or those that come in rapid succession is another. When I speak specifically of stillness in relation to idle time, I am suggesting there is merit in learning to be calmly and fully present in any given moment. Experienced meditators will tell you that this type of stillness comes only with great practice, and that a lack of practice leads to feelings of anxiety and agitation when distractions are unavailable. When practice has occurred, however, and we are able to become still in a given moment, our full selves are able to come to the table. Our bodies may be at motion and our emotions can be activated, but the stillness of the mind can regulate these systems and provide a sense of groundedness, regardless of what we are facing. To develop this ability, however, we must value our idle time as potential-filled and protect it in order to practice the hard work of breaking the distraction habit.

This could be important for several reasons. Self-soothing skills, emotional regulation, critical-thinking capabilities, boredom tolerance, and creativity might all be enhanced by simply putting ourselves in the uncomfortable new space of stillness. Without doing the intentional work of saving some of our idle time to develop such skills, the opportunities for practice elude us and the malicious cycle of stimulation-distraction-information sets in. Not only does this rob us of our ability to practice tolerating stillness, but it also keeps us valuing *being informed* over learning *to be*.

SHOWING OUR WORK

British poet Malcolm Guite refers to the benefits of stillness when he discusses the process of making art. Artistic creation, as he experiences it, slows us down and provides a space wherein we can utilize both reason (that which we can weigh, measure, and observe) and imagination (those things that relate to value, quality, love, and intuition). Together, in Guite's brilliant view, reason and imagination provide two eyes with which to see an integrated and balanced whole. This type of vision requires a stillness, or idleness, within which our attention can be both directed and shaped. Without this stillness, we simply see rather than process that which is seen, and we foreclose on the important work of synthesis and deeper understanding. To illuminate and explain his hypothesis, Guite suggests that in life, as in elementary school math, "showing one's work" is more important than simply penning the correct answer. Absent of the persistence, internal processing, active problem solving and creative invention, and self-direction required to arrive at a sum, the answer means little.

Back in the 1980s, my husband would call the reference desk at the New York Public Library to enlist help in finding information he couldn't locate elsewhere. A reference librarian would listen intently to what was needed, promising to do all necessary research to find an answer and to call back within twenty-four hours. When the librarian returned the call, there was frequently an interactive conversation about the process by which the data had been collected and how the answer itself was more complex than the original question. After a personal pursuit for information followed by at least a day of waiting, the significant investment in the outcome was obvious and rewarded in a way that instant gratification can only hold a candle to.

Today, this process is typically reduced to an online search requiring little engagement with the question or spacious, attentive idleness for processing. We experience a need to have accurate information expediently delivered, so we forgo the kind of engagement to which Guite refers. Along the way, we create a cycle wherein we rob ourselves of opportunities to develop the very skills and abilities we need to think independently, critically, deeply, and creatively about the questions confronting us. We don't practice getting still with our thoughts and feelings, nor do we practice creating space to do (and later show) our work. This makes us more likely to avoid such engagement in the future and leaves us bereft of experiences where we do the work so that we can show it.

When so much of our idle time is used to amass data or digital experiences, our embodied experiences and deep critical-thinking capacities may take a hit. I am profoundly curious about how having so many answers constantly close at hand is impacting our development as humans. Born with everything we need in the way of muscles, organs, flesh, and bones, many of our biological systems require engagement to mature. We are satisfied by milk for a short while after birth and then move on to desiring and requiring more complex sustenance to thrive. Crawling leads to walking, and babbling and mimicking lead, eventually, to talking. Our brains, and the patterns of cognition and affective response housed therein, come to be what they are, at least in part, based upon the type of information and experience to which we allow ourselves exposure. This embodied social and sensory-motor process, discussed in previous chapters, creates the deep grooves in the brain that then guide our responses to life.

According to the neurological principle of "use it or lose it," a range of skills and habits regarding the handling of time and information are either developed or they are not. If we respond to questions or cognitive puzzles by regularly seeking the "correct" or expedient answer, available to us in a Google search, we foreclose on the important work of accessing and developing our critical-thinking skills. Over time, the abilities to engage cognitive processes and wrestle with the unknown become so diminished that they atrophy. This interplay between experience and ability is cyclical. When we develop an ability to wrestle with ideas, data, and concepts, we accomplish a sense of competency with this process. This increases the chances that we will intentionally engage opportunities for inquiry, critical thinking, and creative analysis. This, in turn, will strengthen our abilities in each of these domains. If, however, we lack a sense of cognitive competence, we will be less likely to expose ourselves to opportunities to develop it. This demonstrates how the feeling of incompetence serves as a deterrent to developing the very skills that would help diminish it.

For example, those of us who know how to throw a ball are likely motivated to use this skill, thereby strengthening our ability. Those of us who feel inept at ball handling will likely avoid activities requiring the throwing of balls, thereby robbing ourselves of opportunities to develop comfort and skill at the task. To be sure, we might likely avoid the ball-park altogether. The same is true regarding idle time and showing our inner, cognitive work.

This is not to say that watching a YouTube video to learn a skill or googling for input on how to tackle a difficult daily task is all bad. It isn't. What is important to note, however, is the frequency with which we rely, full-scale, on answers found via our devices without discerning which of these might be answerable by a bit of cognitive wrestling. Neither is all good or bad, but a complete handing off of our cognitive processing skills likely has consequences. So does our tendency to believe that more information is better than less. We would be wise to think this through. Does accessing every piece of data in which we are interested make us smarter and more grounded as humans? Or might we benefit by valuing and practicing the kind of stillness that will help us think deeply, act more creatively, and tolerate the stillness or internal wrestling that creative problem solving requires? Finally, what might be birthed by our *not* having an answer?

BOREDOM TOLERANCE

Boredom tolerance is exactly what it sounds like: an ability to tolerate being bored. While much current research touts the value of this skill for our creativity and self-efficacy, most of us have exceedingly complicated relationships with unaccounted-for time. In our busy and full lives, we speak of craving *down*time, but when it presents itself, we often fill it with little commensurate awareness of our actions. When moments of idleness present themselves, we check in on social networks, catch up on news, play digital games, watch videos, and mindlessly surf the internet. For many of us, using our free time in these ways helps us feel both refreshed and productive. Not only does this indulge our unconscious belief that productivity is better than a lack thereof (e.g., passing through another level of your favorite game makes you feel as though you've accomplished something), but it also leaves us with little time that is truly unaccounted for. At the end of the day, we are incapable of tolerating boredom because we save little time for being bored. In so doing, we shortchange our developmental potential.

Learning to live with open and nondirected time brings with it many gifts. Boredom tolerance correlates positively with measures of creativity, and experiencing intentional boredom paves the way for learning to function in "being" states as opposed to "doing" states. Being states are experienced when we see past all our actions and accomplishments to who we most truly are. From a place of being, we can encounter our most authentic selves and explore and work through our motivations, emotions, beliefs, and worldviews. At our core, we are secure or insecure, confident or shaky, integrated or incohesive, and more. These complex states of our very being end up shaping the way we live and move about in the world. While we may brush up against these qualities in ourselves when we are active, it is only when we are quiet and open that we can truly come to understand and shape these foundational parts of ourselves.

For this reason, an ability to tolerate stillness with or without boredom is important. From here we must also be able to handle the special experience of boredom, because it is a place from which we can experience the "edges" of our personal experiences. When we are bored, we learn our capabilities or lack thereof. We find out how to stimulate or soothe ourselves. We learn to determine and meet our needs from this place. If all we ever meet boredom with is an impulsive action to distract or engage the self in pursuits outside the self, we will never be fully capable of functioning from a space of purely being who we are.

Being states are undervalued in our present hyper-accomplishment and productivity-based culture. Many of us have never been taught to value such experiences, let alone how to find and nurture them in our daily lives. Just as we feel anxiety when faced with a hard math problem, many of us experience a pit in our stomachs, a spinning in our minds, or a stuckness in our hearts when we come across an idle moment in particular or boredom in general. This response to being moments (often experienced as boredom) is, at its core, anxiety, which almost always incites a fight, flight, or freeze reaction. Because of this, boredom tolerance—and its close kin anxiety tolerance—are twin requirements for learning to tolerate stillness.

ANXIETY TOLERANCE

Anxiety is a fierce and sneaky force. It changes shapes and can slip through even the most solid defense. Anxiety and its cousins, restlessness and worry, hang around hypervigilantly, waiting for the moment when we let down, take a break, or breathe deeply. In these moments, they reach into the silence and loudly remind us that all may not be well, there is more to do, and that we "look like total idiots standing around doing nothing." When we haven't taken the time to get to know anxiety, or if we've never been conscious of its presence in our lives, we can easily mistake it for our conscience or for a voice of wisdom and caution. This causes us to act on its message: to get going, to stop standing still, to not look like an idiot, or to do something (anything) to stop the frantic feelings. In this way, we never confront the real cycle directing our thoughts, feelings, and behavior. We never discover that anxiety is not our friend but is, rather, preventing us from developing the kinds of skills that will enable us to live through uncomfortable moments and to find our truest and deepest selves.

Learning to discern what is anxiety and then how to handle it is difficult work. It is, however, mandatory if we want to be able to move through it to places of freedom and potential. Psychologist Elsbeth Martindale describes how even a small amount of tolerance for our anxiety can allow us to "go around the bend," moving away from the behaviors and feelings that reflexively result from our anxiety toward healthier ways of relief or working through. What she is saying is that we need an ability to tolerate feelings of anxiousness in order to learn from them in such a way that we can move away from them effectively. Tolerating our anxiety enough to examine it critically and understand the way it works will take us a long way down the path of being able to engage in new ways of being (think still and, perhaps, bored).

When we have not yet mastered a tolerance of our anxiousness, encountering boredom can be nerve-wracking. The quietness presented leaves ample space for self-doubt and a host of other uncomfortable feelings. Thoughts can be tricky here too. We simply aren't used to letting the mind and emotions sit still. As a result, when we do, they act out. Just as I might respond with a lot of loud internal commentary when faced with a difficult math problem ("I will never get this. I can't believe how incompetent I am. I am so stupid."), which would in turn impact my ability to effectively tackle the problem, most of our minds respond with a flurry of activity when faced with boredom. ("This is boring. What can I do? Seriously! How long do I have to tolerate this nothingness? Aaaahhhhh . . . stimulate me!") Learning how to live through, rather than be diverted by, this noisy self-talk is one of the most important parts of the process of human development. To be our healthiest and sturdiest selves requires an ability to be with ourselves in all our states of being; this enables the cultivation of imagination, engagement with complexities of thought, and a familiarity with our feelings. Although this often externally looks like standing still and might look or feel like laziness, it is in reality much more of an idling where much internal activity is going on, even if the body is still.

HAVING TIME VERSUS MAKING IT

Intolerance, inexperience, and anxiety are not the only things that keep us from engaging idle time to help us grow. Personal habits, value systems, and cultural norms also conspire to keep us from protecting downtime. Everything about our world moves quickly, and most of us experience more demands than we can effectively address. Rare is the person who experiences time as a plentiful resource. Most of us, in fact, would say that we don't have access to any idle time at all. This is a flimsy excuse. It's not that we don't *have* the time to idle; it is, instead, that we do not *create* time for idle spaciousness.

Long ago I decided to stop using the phrase "I don't have time to . . ." and to replace it with "I choose not to make time for. . . ." The conviction I encountered when I made this change was huge. I felt different saying, "I *choose* not to make time for professional and personal development/self-care/relationship investments," than simply spouting off, "I don't have time for. . . ." In truth, I *was* making choices about how I used my time. Sadly, however, I just wasn't making those choices in conscious or intentional ways!

We all do this. Our days are full of moving from space to space and task to task. In between these movements, we have moments we can choose to fill or leave open. Much of the time we make this choice with little to no thought, frittering away the moments of potential that present themselves to us. When we do this, we often find ourselves feeling or speaking as though we are victims of our schedules rather than masters of them.

These moments, in many ways, are like the paperclips of our days. Paperclips are easy to discard, given their perceived lack of value and the sense there will always be plenty of them. The truth is, however, that they offer specific and valuable help when we need to hold papers together. When we save and collect them in mindful ways, we can know we will never have to spend time or resources finding them; they'll simply be there for us. The same is true for idle moments. When we actually notice them, keep track of them and how we spend them, possibly even aggregating them to get a greater impact, we can know that we can count on them to provide opportunities for growth.

I wrote earlier about my Ford Custom 500 and her strong idling motor. Given the uniqueness of everything about this car, she was named (Ethel) and known about town for never being in "park" for very long. I lived a very full life in my high school days, and she effectively delivered me from commitment to commitment quickly and effectively. One day on the way to school, I realized I had forgotten something at home and pulled around the block and up to the curb in front of my house, putting Ethel in "park" and leaving her running as I raced into the house. When I emerged a few minutes later, she was gone. A quick scan found her, still running, backed firmly into my neighbor Pam's tree across the street. Not used to sitting still, and idling high, Ethel had knocked herself right out of "park" and into "reverse."

I can relate to this response to idling and believe that many others can as well. Life, largely bereft of natural opportunities for alert stillness, moves quickly. We have habituated to a furiously fast pace and have not maintained disciplined practice in the art of engaging idle time for the purpose of growth and deepening. Instead, when we find ourselves stopped, we often impulsively knock ourselves into gear, finding movement, distraction, and habits more familiar and comfortable than boredom. Such movement also feels as though it rescues us from the stuck and spinning feeling that anxiety can produce. Perhaps it is time we make a different choice and, unlike Ethel, learn to be still in order to find the resources available to us only in the spaciousness that idle time offers.

8

Cultivating an Internal
Locus of Control

I recently did a series of talks in a small town in northern Oregon. I feel a particular attachment to this area, as my maternal grandparents met, married, and raised their family there. My mom and her sister both reside within its city limits, my uncle is the local celebrity baseball coach, and he and my aunt own the home in which my grandmother was raised alongside her fifteen (yes, fifteen) siblings. As a child, I grew up taking long road trips from central California to this upper edge of Oregon, and it still feels oddly like home to me. Not surprisingly, then, when I was asked if I might consider extending my time to present a kickoff assembly for the high school's upcoming Technology Awareness week, I was excited. The fact that the date coincided with both Mother's Day and my mom's birthday—and that she could be present for the assembly—made the whole situation seem even more perfect.

On the day before my time at the high school, I delivered two lectures for the county employees and community members interested in the topic of how technology is impacting our lives. Both talks were well attended, and the coordinator nearly gushed with gratitude and excitement about how things had gone. The primary lessons I had tried to drive home during this day of talks included:

1. Know your stuff. Stop making all manner of assumptions about things you know little about. Find out what's out there in cyberspace. Play the video games your children play with them. Learn how to tweet and text, and put yourself in the spaces where youth and young adults sometimes live. Stop being shocked and appalled and get on board with becoming an informed navigator and guide for children.
2. Start non-shaming conversations about your own and others' technology use. Listen more. Love more. Seek to understand more than you seek to teach. Get over your wish for how things *might* be and be in touch with how things *are*—especially in the areas of sexuality, vio-

lence, and monetized media/digital platforms. Use this knowledge to help craft empathic, loving, and humble conversations. Stop shaming; it never helps. Listen more. Talk less. Model well.

3. Live wild, fiery, embodied lives and invite others to do the same. Put down your own device. Develop interests and ideas and include others in your pursuits of them. Do what you can to encourage and enable others to put down their devices to engage with embodied living. Find and offer muses worthy of leaving technology for. Don't let technology or the messages that it delivers become more important than your embodied life/understanding.

The county employees and community members responded enthusiastically to my message. Workbooks for helping families make healthy technology use plans flew off my merchandise table. Teachers expressed excitement about adding information I had provided to their curriculum, and parents said they were eager to ask their children to forgive them for talking smack about them without extending the offer to listen and learn. Needless to say, I was thrilled to have helped offer tools and ideas to individuals who were excited to run with them. I also felt deeply glad to have been able to advocate for respectful, empathic, and direct communication with youth and young adult culture and to have people really hear this. Building a bridge between generations, users and nonusers, technophiles and Luddites, is one of the aspects of my work about which I am most passionate, and this particular day was one in which I felt I had advocated well and thoroughly for a young adult population that wants and needs the love and connection of their elders.

As the evening went on, my email inbox filled with questions, requests for more information, and inquiries about where I'd be speaking next. Of special note were the many reactions to the fact that I would be with the high schoolers the next day, presenting brand-new research and data and guiding them through a ten-minute silent sit. This sit has become an important part of all talks I give to adolescents and young adults, as I have found this population eager to step away from their staggeringly distracting devices. I find that they crave to learn how to do this. More than ten people commented to me directly or in an email that they'd love to be flies on the wall watching the teens in their town try this. Equally as many told me I was nuts.

With all this positive energy and excitement fueling me, I did my usual pre-talk late-night perusal of the last week's stats regarding app use, new platforms, and relevant research, updating my slides with data as new as one-day old. I added in video clips specific to students' expressed interest from this very school and borrowed an "I love my FISH" T-shirt (yes, the Astoria High School mascot is the Fisherman) from my cousin so that the students would feel my loyalty and desire to honor their unique school before I even opened my mouth.

For the first twenty-five minutes of the hour-long assembly, things went well. I began by showing a clip from a hip television show and then quickly got on my knees, asking for their forgiveness on behalf of my generation. "We have handed you this amazing thing called technology, and then we have resented you for using it. We have become hyper users ourselves and rarely own that. We tell you to put your devices down but don't teach you how to handle the strong feelings that come up when you do. For these reasons and more, I am sorry. On behalf of my generation, I ask your forgiveness."

I went on to present data about how technology makes our lives better and how it has some detrimental effects on our relationships, sense of self, and brains. I had buy-in from much of the assembled group and had a handful of students who were nodding assertively at even difficult points I made. Then everything went south. My computer, for some unknown reason, froze, making it impossible for me to advance my slides or see my notes. With six hundred high schoolers sitting through a mandatory sixty-minute assembly in a room with no windows on a beautiful day in front of me, I had to punt. I had them stand up and act out a living brain. Without my computer, I struggled to remember the sequence of material I had so carefully lined up the weeks before and the research I had carefully evaluated and added the prior day. I recalled what I could and got us to our final ten minutes, when I led the students through a mindfulness meditation followed by seven minutes of silence.

By the time the assembly ended, I was a sweaty mess. It's never easy to hold the attention of large groups of adolescents, but to do so with no notes, no technology (I am heavily dependent on well-placed videos, music, and interesting images in my talks), and a silent ending felt like a feat of epic proportions. The students who had questions about specific research, and those who felt excited about the challenge I had offered of doing ten minutes of mindfulness a day, rushed the stage to chat with me afterward. Even more students, those who resonated with some of the ideas I presented or wanted to exert well-formulated challenges to my research, hung in the wings and talked with me after the auditorium cleared. But the students who disliked me, who heard messages in my talk that disturbed, offended, or upset them, waited until later to approach me—on Twitter.

To be fair, I invite people to tweet during my talks. I have had many instructive interchanges there and have had folks make important suggestions and corrections in that space. When done with respect and care, all is well, and these communiques refine and inform my work. I accept this kind of criticism with difficulty and commensurate gratitude. But the comments of these particular students were not carried out in this way and were instead lobs at my message and at me. One rated my talk a "10 on the dumb meter." Another asked me to never return to the school or city again, as I had made "a fool" of myself. Comments I made about self-injury and ADHD were "quoted" inaccurately and taken out of context, making me sound ill-informed, off-putting, and uncaring. And, who dressed me anyway? I was a "fat, middle aged, clueless poser who flamed and shamed everyone with ADD or depression." Nothing was off limits in these messages, and as students began noticing them, the comments and messages followed.

As the tweets started pouring in, I felt myself becoming deeply agitated. The floodgates of self-doubt were opened, and before long I was unable to think about anything other than what a terrible mess I'd made that morning. With no recall of anything positive that had occurred before those tweets, I began to doubt not only the quality of that mornings' message but also myself and my abilities. I was talking with my mom, appearing to enjoy a leisurely meal, chatting about a few shared events, and trying to celebrate her and ignore my exploding phone. Inside, however, I was both beating myself up and trying to figure out how to manage the situation playing itself out online. I couldn't focus or be truly present, given the cacophony of internal dialogue bantering about in my head.

Feelings relating to what these students thought of me began to take over any sense of reason or rationality. Soon, all my talks from the past twenty-four hours were recast in my mind as total failures. I began to add up the skeptical looks from the prior day and the questioning glances and giggles from that morning at the high school. I reconsidered the requests for more information and for references of data that I presented, deciding that most of the people I'd been with those two days wanted to do their own fact-checking and expose me as a lying, non-empathetic, intellectually challenged, doomsday, self-promoting, shaming fraud. Never mind the few folks (I had reduced those who spoke of benefiting from the presentations to a very small group) who benefited! Clearly they were either lying to spare me embarrassment or clueless regarding quality and content.

Later, as I crossed the city limits headed for home, I realized I had taken on a fully unconscious belief that I was leaving the city disgraced, having presented information that was both inaccurate and insensitive. In my gut, I truly felt as though I were being run out of town. I pulled off the road and sobbed. I had lost touch with my own sense of certainty and clarity around my work and had given my mind and feelings over to the world outside of me. So strong was my buy-in to the opinions of others that I didn't even question the credibility or intention of the sources—and I study and research this stuff! I was convinced I needed to retire.

This situation presents a small microcosmic view of what can happen for all of us every day. There is our own internal experience of an event or experience—and then there is the external world's response to or interpretation of it. This observational reality can be applied to both ideas and actions as well as to our very personhood. We might experience support or praise from outside ourselves for a certain thought or action we have taken. We can also, however, receive support or praise simply for being. Negative feedback functions similarly and can be offered based on who we are or what we think or do.

Beginning early in our lives and continuing through our development into adulthood, patterns are created that influence the way in which we make sense of our internal and external experiences. These patterns are shaped by the kinds of experiences to which we are exposed as well as by uniquely personal factors such as temperament, the culture and families in which we are raised, the goodness of fit (or lack thereof) with our primary caregivers and communities, and the relative ease or difficulty of our past and present circumstances. At the same time these proclivities are developing, we are establishing important relationships with our own selves and others. The nature and strength of both our self-self and self-other relationships determines much about our health, groundedness, and stability in the face of both everyday circumstances and highly stressful ones.

Basically, the center (loci or locus) of control regarding how we feel about ourselves and about the world can either reside inside of us in our most grounded core or outside of us in our community or the world. The individual who tends to look to sources outside him- or herself to determine how to think, feel, or be is functioning with an *external* locus of control. Conversely, the person who can turn inward to determine how to act and to determine what he or she is thinking or feeling can be considered to have an *internal* locus of control.

Externally focused people are frequently hypervigilant about the facial expressions, tone of voice, and comments of others. They are overly aware of what "should" be and which thoughts, feelings, and actions will win the approval and acceptance of the majority crowd (whichever majority crowd they choose). They are frequently more familiar and comfortable with *action and doing* and less comfortable with *being*.

On the other hand, people who live from an internal locus of control are more likely to evaluate situations and stimuli based on their values and gut long before looking to anyone else for their opinion. They are comfortable with silence, familiar with a wide range of feelings, and demonstrate flexibility without compromising what is important to them. They are not oppositional and yet are also not afraid of disagreements. They are willing to be different and yet don't need the attention of being seen as such.

These are individuals who others often describe as "grounded." They are resilient and can handle rejection, disappointment, and awkwardness, and they take appropriate risks because they have what it takes to fail.

DEVELOPMENT OF A LOCUS OF CONTROL

When we are born, we cannot independently care for ourselves physically or emotionally and are completely reliant on the people who have taken responsibility for meeting our needs. Whereas we are whole people at these young ages, we do not know how to discern our needs and meet them, making us reliant on forces outside ourselves. If the people available to us at this point respond to us in consistent, reliable, and available ways, we internalize a sense of security in our attachments, believing that our needs will be addressed appropriately. The nature and the quality of these attachments allow for the creation of a template of sorts for what it looks like to become aware of and responsive to our needs. If the attachments are healthy and consistent, we learn from them how to address our needs and, over time, begin to take over the process of meeting our own needs, using their actions on our behalf as teachers. If the attachments available to us are not healthy or consistent, we either continue to seek external sources of validation and care in the hopes of eventually procuring good-enough modeling or develop a rough, rigid individualism that cobbles together ways of managing the care of self based on whatever we can find.

Playgrounds offer a great analogy for this discussion. Imagine a young child visiting a local park with a caregiving adult in tow. If the child falls and an injury results, the child commonly will look immediately to the caregiver, often demonstrating no sign of an emotional response until doing so. Typically, one of three responses from the accompanying adult commences:

1. The caregiver reacts strongly on one end of a "this is a huge deal" to "this is not a big deal at all" continuum. This is the caregiver who either diminishes or overreacts to the incident, drawing either too much or too little attention to what has occurred, without actually assessing both the physical and emotional impact of the child's experience. A response of this sort communicates to the child that it is the caregiver's assessment and style of responding, more than the child's own assessment, that matters. The child in this scenario comes to rely on the reaction of the caregiver—whether that reaction fits the situation or not. Whether the caregiver responded with an externally dismissive response set (e.g., "My caregiver tended to underreact when I looked toward him.") or an overly reactive one (e.g., "My caregiver repeatedly overreacted regardless of what had really occurred."), the child habitually interprets his or her situation from a lens *outside* his or her own center.

2. The caregiver, whether physically present or not, responds with complacency or absence. This response leaves the child to deal with the situation independently and sends a subtle message that there will be no help to deal with the consequences of life. The child in this scenario has a proclivity for developing either (a) a sense of rugged independence that excludes honest assessment of how he or she genuinely is or his or her desire for connection or (b) a voracious hunger for the outside world to meet his or her needs.

3. The caregiver responds to the child's inquisitive glance with an empathic and connected retort calling for assessment. "How are you? Are you okay? How can I help? Let's figure out what has happened here." This partnering response communicates that the child's own assessment of and communication about the situation is crucial. This response puts the child in the driver's seat by enabling him or her to slow down and consider what he or she needs and then allows for a partnership in addressing the need. This response also communicates that, ultimately, only the child can know the initial level of injury/pain/disruption, yet there is a person with a bigger toolbox ready to help remedy the situation once it has been assessed. This caregiver uses the initial, right-after-the-fall glance from the child as an opportunity to loop the child's attention back to the child's present state to determine the severity of the situation from the cues the child's own body and mind give him or her. This kind of response helps the child recognize and make sense of his or her own realities—before looking *outside* him- or herself.

While this exploration of response style is overly simplified, the frequency and pattern of responses we receive from our external support systems while growing up go a long way toward determining how we will react to missteps and falls of all types over the course of our lives. By late childhood, we typically have come to respond to circumstances and experiences in a unique-to-us, routine way. If we have developed an ability to assess destabilizing experiences, determining what we need to resolve a situation or comfort ourself, we function from a primarily *internal* locus of control. If instead we tend to base our feelings of being okay in the world on what others think about us and what we should do or how we should act, then we function from a primarily *external* locus of control.

The tendency toward either an internal or external locus of control involves much more than just our upbringing. Regardless of the response patterns of our caregivers, our own personalities and temperaments shape our proclivities, as do the circumstances we have faced. In addition, even for those of us who had caregivers who did their best to help us learn to know our minds and hearts and to live healthily from them are influenced by a culture that rewards an external locus of control.

In the United States, a vast proportion of our lives have been lived in highly competitive environments that thrive on us being constantly aware of where we stand in relation to others. This culture of competition encourages a constant scanning of the environment to learn what will best achieve a one-up position. Such a culture makes us vulnerable to caring more about reinforcement from the outside than about living from an integrated center. To get ahead in this kind of majority culture, we must constantly be aware of which actions, accomplishments, and acquisitions will secure a more desirous place for us in the world. Making matters worse, the fact that we have round-the-clock access to connected platforms means that we have more potential than ever to look outside ourselves to determine how to feel, think, and act. This leads to a consistent temptation to forgo the knowledge of our internal world in deference to seeking the rewards of our external ones.

In a capitalistic culture, there are benefits to majority culture maintaining an external locus of control. It benefits the corporate-minded West to perpetuate a culture of comparison and competition. When the GPS in a phone allows businesses to push notifications to consumers in the proximity of their products or services, it feeds the notion that one's device can discern and meet one's needs apart from deliberate personal thought or action. When a population relies on constantly available advertising and crowd-sourced reviews regarding what to purchase and consume, a market-driven economy will thrive. The ever-present message that there is more we can buy/do/accomplish to raise our social or personal status is deafening—and a strong but subtle competitiveness thrives in digital spaces where a culture of public comment, responding, and assessment keeps us aware of our standing.

When we turn to our devices more habitually than to our own values and preferences, we leave ourselves vulnerable to living from the outside in. When faced with good news or bad, we turn to online communities for help with how to interpret. Feeling uncertain as to how we measure up, we scan the profiles of others and judge ourselves against their norms. That which we see online as valuable or meaningful becomes the bar by which we measure ourselves. Bereft of the ability (or time or energy or more) to withdraw into ourselves to assess, comfort, and determine action, we look to our devices and digital spaces to both comfort and motivate us. This is living from an outside-in, external locus of control.

The truth is that developing and maintaining a life lived from the inside out is not easy. When we have access to a never-ending stream of information, opinions, ideas, and stimulation, it can be harder work to know oneself than to rely on the collective knowledge of the internet and the communities it delivers us. In fact, never before has it been so easy to substitute an external locus of control for an internal one. Feel overwhelmed, bored, or insecure? Find everything you need for entertainment and information on your device! Not sure how to feel about yourself? Look online!

This is exactly what I did when faced with the negative tweets after my talk. Unable to get myself quiet enough to determine how much weight to give the comments, I simply believed them. In fact, I inflated their significance, given that they lined up with real fears about my intellect and that they existed in an online space where others could see. Once I had given these external sources power over my feelings about myself, I looked for evidence. I could have just as easily turned to my external communities to buttress my sagging self-esteem. Instead, if I would have lived from an internal locus of control, I would have gotten quiet and still, validated the feelings of hurt over having been publicly called out, then determined what I needed to move forward. I might have thought about other times I've faced feelings of disappointment and discouragement and determined what has helped me to address and resolve these feelings. I basically would have *responded* to the situation rather than wildly *reacting* to it, and likely would have experienced much less upset.

All of us, I believe, could benefit from developing a greater internalization of our locus of control in relation to our thoughts and feelings about, as well as our reactions to, life. This way of being is more about living from an examined core. It is about self-love and self-respect rather than self-promotion. It is about knowing ourselves honestly, recognizing that others exist as whole beings, not just as external sources of comfort, stimulation, and affirmation. This way of being in the world allows us to live in radically meaningful interdependent relationships, demonstrating an ability to be both alone and with others, to function as either leaders or team members, and to act according to our own values without diminishing or judging those whose values are radically different. We can feel a sense of groundedness in people like this, people who march to the beat of their own drums without needing to silence anyone else's beat.

People who have developed a strong internal locus of control are often seen as countercultural and stand out in a crowd. They know their strengths and their weaknesses and do not over- or under-account for either. They are careful without being cautious. They are humble without being false. They are steady without being rigid. They know and own their opinions and biases and can see past them. They are not intimidated by other strong people and are neither threatened nor threatening when among others who are different from or disagree with them.

An intentionally developed internal locus of control allows a person the ability to approach each human encounter with the perspective of leaving that encounter having respected one's self as well as the other (whether that be the other person, idea, experience, etc.). In fact, a hallmark trait of individuals with a solid sense of self is that they do not approach the world with the intent of overtaking it; they simply want to contribute to it. They want to make an impact toward the direction of greater health, integrity, or authenticity. The goal for these individuals is not to change the mind of another or to convert or win over another. Instead, their goal is to encounter each person in a way that respects and acknowledges that individual's personhood while, at the same time, not shying away from one's own deep convictions. When these folks disagree with someone, a respectful space for a differing of opinions opens up. At the same time, there is a basic honoring of the dignity of the personhood of the other.

Having a compassionate, empathetic internal locus of control is an exceedingly advanced skill to master, as humans are much more likely to operate from an us-versus-them or narcissistic mentality. Humility and resilience are required to function in a way that does not trigger defensiveness or reactivity when presented with differences. In addition, we often surround ourselves with people like us, which robs us of the opportunity to practice the necessary skills and develop the ability to tolerate differences. An internal locus of control requires both the opportunities to develop these skills—and the courage to lean into them.

Modern life does not automatically offer opportunities to live from this integrated center and instead bombards us with input twenty-four hours a day. We are never far from sources vying for the opportunity to inform us on how to think about ourselves. Constant streams of advertising, push notifications, and on-screen images that stare out at us from every corner feed us a never-ending host of aspirations and wants. The easy access to platforms that allow us to solicit the opinions, responses, and validations of others throughout our day doesn't help. It all blends together to send the message that if we are just "in tune" and comply with what *appears* to be best, we've "arrived."

Living this way comes at a profound cost. Within a culture that offers and reinforces an external locus of control, self-promotion becomes a way of getting what we need. No longer is the community, the collective, the group, or the team the focus. Rather, the collective, the group, and the team exist to validate, reinforce, or even promote the self. The focus morphs from "How can I be a part of this collective?" to "How does being a part of this collective help, reflect on, or impact me?" Whereas this may seem like a subtle, indiscernible difference, it has profound implications.

SELF-PROMOTION AS A PRECURSOR TO
AN EXTERNAL LOCUS OF CONTROL

I already mentioned my "celebrity" uncle who is a well-decorated high school baseball coach. Over the course of his thirty-five-year career, he nurtured hundreds of players individually and led many teams to state championships. In his early years of coaching, the process of team building was relatively easy. Culturally dictated norms enforced the thinking that a healthy and well-working team was more important to the high school athletic experience than was promoting the personal profiles or prowess of individual players. This meant that, more often than not, players joined teams ready to forgo the personal limelight in order to work as part of a well-functioning group.

As time went on and even young children got sucked in to participating year-round on club teams in a single sport (as opposed to playing multiple sports, each confined to a specific season), my uncle noticed that his players began to feel a sense of pressure to give up involvement in other activities in order to devote increasing amounts of time to their school and club baseball teams. As their investment in their baseball skills became greater, the well-roundedness of these athletes who previously played a variety of sports, and possibly excelled in other pursuits as well, decreased. In addition, the stakes rose regarding the potential payoff, or lack thereof, of their dedication. Vying for limited spots on teams meant sharpening the competitive edge for the individual, which came at the cost of team development.

In his later years as a coach, my uncle noted that most of his players had private coaches in the stands. Those players would often look to their private coaches rather than to my uncle, the coach of their team, during competition. This subtle shift toward emphasizing individual play over team play is significant, especially when the entire team has made the shift and each person is playing for his or her own rewards rather than for the reward of playing together.

My uncle's experience is shared by many who have had long coaching careers in the United States. In fact, popular culture in general reflects this progression toward self-promotion in spades. It isn't limited to the athletic arena but thrives also in business, the arts, and academia. As fallout of this need to self-promote, studies show that high school students are experiencing increased levels of anxiety, as early as freshman year, about what they need to do, both in and out of school, to gear their transcripts toward college admission. Parents of children in early elementary school are receiving letters from high school coaches and extracurricular leaders, informing them that their children may not earn a spot on the school's team/club/etc. unless they have specific, advanced training before entering high school. The kind of admissions competition that used to be limited to the most elite universities is now even found as parents try to secure a spot in preschool for their toddlers!

With so much of our "success" being dependent on our ability to navigate and work within the current cultural economy, it's not surprising that many of us lose track of what is healthiest and function instead from a strongly external locus of control. Most commonly, we develop that external locus of control because we either don't trust or know how to access our own leanings, haven't learned to speak or hear our own minds and hearts, or, quite literally, have had no models of how to nurture a trusting relationship with our unique selves. We know there is an out-of-sight, out-of-mind reality to our social circles and that those same circles live in social networks, so we engage there to maintain our social standing. We are aware that admissions officers, potential employers, and future dates will scour our online presences, so we tailor them to accomplish certain ends. We recognize the importance of building a personal brand, so we bend the truth here and there to cater to a larger portion of the population we hope to attract. All the while, we know we are not being fully authentic or congruent—and that each follow or like is being fed a version of ourselves that, if that person met us in embodied life, he or she might not follow, like, or even recognize. In fact, we chose self-promotion over honest expression in a million large and small ways.

Part of the difficulty of this approach to life lies in the fact that the more rewarded we are as a result of these promotional outreaches, the less experience we garner in developing an authentic self and living from that self in the world. We feel uncertain of who we are and unsure of how to develop ourselves. This leads us to feel vulnerable in reaching out to others, making us less likely to do so. We feel lonely or alone as a result, leading us to find solace, support, or connection in digital spaces where it feels easier and a bit removed to take relational risks and where we can put emphasis on certain desirable aspects of ourselves and downplay those we fear might be less acceptable. This negative cycle perpetuates itself, leaving us less and less familiar with our authentic selves and increasingly comfortable with our external worlds dictating how we feel about ourselves.

Overuse and excessive reliance on technology feeds this cycle. We lack practice in relying on our own selves to know how to comfort, stimulate, entertain, and soothe ourselves, so when we face boredom, insecurity, failure, or other strong feelings, we look to either our devices or the connections within our digital spheres to give us almost immediate feedback. We are alone at a restaurant or in line and feel self-conscious that we have no one with whom to talk, so we look to our phones. We face an evening with no plans and feel incompetent in the face of unaccounted-for time, so we binge-watch Netflix. We experience insecurity when we haven't had a meaningful conversation for a while, so we fish for someone with whom to text just to fill space. We lack affirmation, so we post our every success as well as the successes of our child or our pet. And then we obsessively check for likes and responses. We measure our "mattering" in this world by the number of retweets or followers and panic when those numbers don't budge. We feel stirred up and have no idea how to release, so we watch a violent movie, play a fast-moving video game, or dial in some porn. If we're really amped up, we do all three.

SELF-KNOWING AWARENESS AS A PRECURSOR OF AN INTERNAL LOCUS OF CONTROL

An alternative to self-promotion is self-knowing awareness, or self-love. When we pursue a deep knowledge of our authentic selves or a sense that we can direct our own growth and functioning, we have the potential to foster an awareness of our strengths and weaknesses. When we learn to hold these insights in nonjudgmental ways, we create an internal economy wherein we develop ourselves based on what is healthy for us personally as well as what is healthy for those with whom we are connected. As we do this, we learn to trust ourselves with how to evaluate what is going on in and around us. We become, in essence, securely attached to our own selves, relying on that attachment to help us tolerate both successes and failures, the known and the unknown. We learn how to know and work with our feelings, our behavioral proclivities, and our thoughts. We become flexible and resilient and able to comfort and challenge ourselves, while at the same time recognizing and honoring the selves of others.

To get to this place, we must be willing to do the difficult work of disengaging from the places we have defaulted to for our external loci of control. This typically requires a long, hard look at how we have relied on our external environments and relationships to fill in for us in determining our actions, thoughts, and feelings. The type of serious self-evaluation to which I am referring here is not a self-centered narcissistic pursuit. It involves instead the work of taking an honest appraisal of how we have offloaded the tasks of self-soothing, of generating thoughts and well-considered worldviews, and of self-correction and affirmation onto others —either those we unconsciously trust and gravitate toward or those we imagine are better at those things than we ourselves are. It requires a careful look at the ways we have overrelied on culture, family, friends, or social/civic/political/religious/professional groups to tell us how we should act, think, and feel. Equally important to note is that if both our sense of self and our relationships with others are built on this kind of dependent need for input, direction, and feedback, we can be sure that at least some of those relationships are built on norms of dependence rather than on healthy norms. Also, as we live in this dependent way, our relationship with our own self suffers because we do not invest in it enough to know it well and to allow it to guide our responses to life.

When most individuals within a culture function in this way, the entire culture is impacted. Although it may seem counterintuitive, when we come to know and value our own unique way of being in the world, we also develop an inherent tendency to seek, value, and respect the uniqueness of others. Rather than relying on others simply to gratify our own needs for guidance, direction, and validation, we see others as the unique people they truly are, and we make space for their unique contributions to the world. This opens the door to connection that is deeply about relatedness rather than about simply using others to fill in for our lack of comfort with our own selves.

MOVING FROM AN EXTERNAL TO AN INTERNAL LOCUS OF CONTROL

If we have existed primarily on the external locus of control side of the spectrum, a move toward an internal one can sometimes feel (and look) like a huge narcissistic shift. Just as a pendulum does not swing to the middle after having been held to one side, a tendency toward overreaction may commence as we break free of external reinforcers. Questions such as "What do I want?" "What do I think?" and "How do I feel?" can feel strangely self-centered when we have been living based on what everyone else wants and feels. Likely, we will need a grace period and strong guidance as we move from placing our feelings about ourselves in the hands of others to discerning and working with those feelings ourselves. Similarly, as we break free from using our devices and the digital spaces they offer as our sole source of stimulation and soothing, we will need help in determining how to ground these functions in ourselves.

This process of centering our everyday living in our core requires that we each have a healthy, honest, reflective relationship with our own self. I propose that the following traits, actions, and capacities are included in the healthy relationship with one's self:

- Capacity for honest awareness of strengths and weaknesses.
- Knowledge of one's emotional range and an ability to moderate and regulate affect emotions.

• Flexibility in relation to the knowledge and recognition as well as the importance of one's personal needs versus the needs of the communities in which someone lives.

• Ability to function independently and interdependently and to foster intimate relationships with others without compromising or disowning important parts of the self.

• General awareness of one's physiological being and ability to be comfortable in one's skin.

The stronger our authentic relationship is with our own self, the more flexible, resilient, and available we are to live lives that are fiery and rich. We are also more capable of maintaining healthy relationships that honor the uniqueness of the other. When we know ourselves and know what works to comfort ourselves, we can begin to take healthy relational and personal risks. We can take actions that stretch us, and we can be poised to handle the rejections and failures that are an inevitable part of life. When we live from our core, we can redeem those experiences that are less than optimal by using them as tools for learning and growth. This kind of living begets a richness and groundedness that is contagious and extremely rewarding, regardless of where that living takes place.

TAKE ACTION

Ideas for Establishing and Maintaining a Healthy Relationship with the Self

LEARN TO BE ALONE.

Start small and go big. Build small chunks of time into your daily routine when you simply sit in quiet and let your mind wander. Grow these segments over time. Go for a walk or hike by yourself. Work up to a movie or shopping trip alone. When you feel ready, go out for dinner by yourself. If you absolutely can't handle the thought of the entire meal alone, take a book. Only read it for a part of the time. At least once or twice, look up and around at everything around you. Take it all in.

Use at least a small amount of alone time to consider your preferences, strengths, weaknesses, and biases.

Do a simple "brain dump" (described earlier) regarding each of these areas: preferences, strengths, weaknesses, and biases. Write for ten minutes straight without thinking. At the end, look for themes. Try to hold the information loosely and be curious about it. What are some weaknesses or strengths that surprise you? What are some you could enhance or live without? Become intentional about how to address the realities you'd like to change.

Talk to yourself or journal.

Get yourself to a place where you can converse with yourself as you would a friend. Ask yourself questions about who you are, what you are comfortable with about yourself, and what you don't know how to access or know about yourself. Explore this conversationally with yourself or journal about it. Come back to this exercise regularly.

Seek out wise others who can help you know yourself more deeply.

We are fortunate to live in a time when there are many resources to help us know ourselves. Psychologists, psychiatrists, therapists, counselors, pastors, rabbis, imams, wise friends, elders within our communities, and more can play a part in helping us know our own strengths, weaknesses, and proclivities. For those who cannot find these wise elders and professionals, assessments might be helpful. When using assessments found online, make sure they are reliable and trustworthy, and always take the findings as suggestions rather than solid "findings."

Part III

Telling Ourselves the Truth and Finding a New Balance

9

Habits and Norms

It is easier to establish healthy norms than to break unhealthy habits.
It is easy to establish habits.
It is not easy to establish healthy norms.
This entire book could be summed up by these three sentences.

Joe is a college senior. He came to the United States on a student visa and enrolled in an East Coast liberal arts college. Early in his college experience, he made note of his classmates' obsessions with their devices and found it difficult to connect with individuals one on one. Determined to do so, however, he left his cell phone in his dorm room during the first semester to force himself to meet people in his new school. Shocked by the inclusion and integration of devices into actual classroom learning and lecture time, Joe quickly adapted by keeping his phone in his backpack in case he was called on to use it. After a full semester of entering the dining hall to tables filled with students looking down at their phones, Joe succumbed and began carrying his phone in his pocket, pulling it out rather than eating alone. When he'd have windows to go off campus, he had a difficult time finding someone who could drive him and couldn't locate a mass transit system close enough to rely on. He decided to download a car-sharing app and used Google Maps to get wherever he needed. He then started bringing his phone to bed with him, eventually tossing the alarm clock he had brought from home. He eventually networked online with other gamers on campus and played video games with them from the comfort of their separate dorm rooms. Over time, Joe lost connection with his intentions about social interactions and evolved a habit of high engagement with technology as a way of making his way in the world.

This kind of progression into habitual behavior is common for most of us. Any kind of repeated action, thought, or feeling can become a norm or habit that can, in turn, create the behavior patterns by which we live and move about in the world. Habits and norms can exist either within our conscious experience or fully outside our awareness, and each of our habits and norms is supported by neurological wiring and behavioral predilection. For the purposes of a shared vocabulary, in these pages "norms" will refer to actions and behavioral patterns that result from *intentionally chosen* goals and considered ways of reaching those goals. "Habits," on the other hand, will refer to patterns of thought, behavior, and emotion that come into existence via repeated actions, thoughts, or feelings that lie primarily *outside one's consciousness*. Habits can emerge from and be related to the norms we set and vice versa. For now, however, we will work with the two constructs as separate but related.

HABITS

All of us are familiar with habits. Most of us have one or two. Habits feel almost instinctual and are often enacted with little or no thought. Repeated actions that seem to happen almost outside our awareness become the habits that shape our lives. The way in which we automatically reach for something in the refrigerator when we're bored, the feeling that comes over us when we make a mistake, or our emotional and behavioral reactions to an ideology that differs from our own are examples of habituated behaviors. Sometimes habits have relatively small and insignificant impact. How we tie our shoes, respond to a ringing phone, or butter our toast might not matter much. The habituated way in which we care for our teeth, however, can have a significant impact on our health. This is true in far more instances than we'd like to think. Because they often grow out of our *least examined* or conscious thoughts, feelings, and experiences, habits are particularly difficult to recognize or change and can profoundly impact the trajectory of our development.

Consider our response to the experience of physical exhaustion as an example. When our bodies send signals that they are tired, a wealth of possible responses exist. If we value rest, we may habitually respond to exhaustion with restorative measures. If we have ascribed to negative beliefs about tiredness, we might habitually shame ourselves for feeling weary or attempt to push past the tiredness altogether. If we are out of touch with our bodies, we may create habits that block any awareness of tiredness from our consciousness simply by rarely being still enough to register it. If we are especially sensitive kinesthetically, we may experience a habituated inflation of our feelings about being spent. All of these responses have their roots in a complex soup of values, beliefs, and conscious and unconscious drives.

The very difficulty with habits is this profound complexity. What seems like a basic automatic action is rarely simple. The way our parents responded to certain experiences, the beliefs and behaviors we learned from our communities, the range of options presented to us over the course of our lives, and more shape the automatic reactions we initiate throughout our days. If direct experiences don't shape the support system of our habits, very likely our reactions to such experiences do. In either scenario, habits have deep roots that profoundly affect our development.

Mature, established plants have root systems that are not easy to excavate. They do the work of supporting their above-earth growth by reaching deep and wide for nutrients, weaving their way into the soil in which they live. Weeds are especially effective at producing complex root systems. When weeds live alongside desired plants, they pose a risk by leaching off important nutrients from the soil that could ideally serve the preferred plant. Weeding is a way of stewarding limited resources by removing competition. When the roots of the desired plant are relieved of having to share water and nutrients with the roots of weeds, there is greater certainty of healthy growth. This only works, however, if the weed's root system is removed in the process. Simply cutting the weed off at the dirt and leaving the root structure intact isn't enough to keep the competing plant from growing back. In this scenario, a small bit of water and light would bring the weed back with a vengeance.

Our habits have deep roots, and the process of excavating these is similar to the process just described for eliminating weeds. Simply choosing to stop behaving in habituated ways is rarely enough to break a habit. Similarly, applying behavioral interventions divorced from work to excavate the deeper motivations undergirding our habits often fails. Each of these actions can easily feel like cutting the habit off at the dirt, leaving its root causes and motivations intact and ready to burst forth in new ways at the first exposure to triggers.

We do this all the time. Let's use habits related to our physiques as an example. Unhappy with our bodies, we frequently expect radical changes from diets or exercise regimens without ever exploring the complicated mix of genetics, lifestyle, environmental, and other factors undergirding the way our bodies came to be how they are. This kind of exploration could yield important information not only about what kind of changes might be possible but also about what might be realistic or healthy. Competing for cognitive and behavioral space with these sources of important information are cultural and contextual messages that subtly shape our wishes and expectations about our bodies. Never-ending streams of Photoshopped and carefully curated images, masses of eating plans and fad diets, and a culture fed on competition around looks and the belief that achieving certain appearance standards can garner us power, attachment, and status live like weeds alongside any thoughts we have about ourselves and threaten to rob us of the ability to think with reason and clarity about the condition of our bodies and the wishes we have for them.

In this analogy, exploring the root causes of how our bodies are and what we might realistically expect of them is akin to nurturing a healthy plant. The exploration of what a plant needs to thrive and grow optimally is accomplished by understanding what that plant is really meant to uniquely look like and what its expected growth and health require. Cultural expectations represent the weeds. They compete for our headspace, so to speak, interrupting information about healthy and realistic expectations and trying to replace that generative information with images and ideas that may or may not be healthy, relevant, or personally applicable. When these less healthy motivators are perceived to include information that will get us ahead in life, provide greater access to a valued commodity (power, esteem, connection), or get more resources (time for thought and consideration), they grow stronger and threaten to choke the growth of the desired plant.

It is hard work to turn our attention from the weeds and to instead nurture the growth of a plant whose characteristics we have limited control over. The near-constant din of cultural noise also serves as a weed in competition for time and attention, sending messages about what will garner us power, love, attachment, and more. This in turn feeds the emotional component of our behavioral habits.

As an example, I can share a personal history. I attended my local public elementary school. While I had a mostly positive experience, I was the subject of some cruel bullying and became fearful about what I might face at a large junior high. As a result, I asked my parents if they might be willing to consider sending me to a smaller school in a neighboring town for junior and senior high. Ever supportive, they made it happen.

I had never doubted my academic skills to this point and was completely taken aback as I began to realize that I was not fully prepared for the academic rigors of my new school. I made friends relatively easily and forged fast connections with my teachers, but I never felt as though I measured up academically. I worked and studied and worked and studied and never achieved the grades for which I longed. Over time, I developed the automatic habit of evaluating myself and my abilities in relation to the context of my classmates—and always found myself wanting. The ever-elusive 4.0 became my obsession. I would stay up far too late and spend every free minute studying, trying to memorize everything I needed for a test or presentation. This would leave me too tired and overwhelmed to actually internalize the material; thus the A would elude me, and a B felt like utter failure. This created a feeling and belief that I simply had not worked hard enough. Even if I received an A-minus, I would sink to feelings of inadequacy and promise myself I would work harder next time.

Eventually, I habituated my life to one wherein I filled every available waking moment with study. I developed a habit of never *not* working. This habit followed me to college and to graduate school, even though I found myself excelling in these settings. By the time I was graduating from college, my habituated feelings of not being enough could not be touched even by stellar grades. I felt as though I was always missing something and that there was always more I could master, more I could know, or better ways of showing it. There was constantly some kind of A out there that I wasn't getting! I came to believe that this was simply a form of self-motivation that would enable me to achieve eventual "success," even though, for all intents and purposes, I was achieving it already. Even external evidence was powerless in the face of this habit of thought—a habit of thought I abhor and with which I still struggle today.

This is the problem with habits: they make themselves at home and get us to believe that we can't function without them. And this is not just a behavioral reality. It's a cognitive and neurological one as well, because our brains are both the product and producers of our habits. This is why changing our eating patterns or the way we spend money or time or energy is so difficult. Habits are hard to break!

Here's how the process works: Neurological function occurs as a result of electrical stimulation in the brain. Our bodies experience a sensory stimulus that feeds the brain an experience. This tips off a string of electrical impulses that travel from synapse to synapse, stimulating the release of neurotransmitters and also creating a groove in the gray matter of the brain. Every time one of these pathways is repeated, the groove becomes deeper and more pronounced. A person smells pizza, the brain interprets the smell as pleasant, and that interpretation trips off many habituated responses, including hunger and memories of times when pizza has been consumed. However, when a person experiences an unfamiliar olfactory sensation, there is no habituated reaction to trip off a synaptic pathway. New interpretive and experiential grooves need to be made in the brain to accommodate the new smell. By this action, our brains become complex highways of deep and not-as-deep grooves. The stronger a habituated response, behavior, thought, or feeling, the deeper the neurological grooves related to it. I had deep grooves related to thinking I was not good enough—and very shallow ones to help me experience or tend to any data to the contrary. Traveling the deep grooves is the path of least resistance.

That is not to say that habits are inherently good or bad. They do, however, direct much of our experience in the world. Because of this, it is important that their presence be brought into our conscious awareness, where we can understand, evaluate, and alter them.

NORMS

An alternative to living habitually is living from intentionally chosen and established norms. The word "norms" is a shortened reference to the phrase "normative behaviors/patterns" and refers to intentionally chosen behaviors and thought/feeling responses that both result and emerge from the conscious creation of healthy patterns. In reality, norms often form the beginning of many of our habits, especially those habits we develop by acts of our will.

Norms are interesting creatures. Like habits, they direct behavior, thoughts, and feelings; but unlike habits, they are established based on chosen, often aspirational, values or goals. The setting of norms involves seeing a possible desired outcome and working backward to determine what is necessary to achieve it. In other words, for the purposes of this book, whereas habits evolve, norms ideally are set down intentionally or are built.

Norm setting is tricky. It involves allowing one's mind to wander from the comfort of what "is" to the unknown of what "might be." It has its roots in finding aspirational qualities, values, thoughts, feelings, and behaviors that might benefit one's life and then finding embodied ways of moving toward those qualities and traits. Norm setting gets at the roots of our behaviors, feelings, and thoughts. When we establish norms, we do so because we understand, at a deep level, that we must make an effort to shape the way in which we are living as opposed to simply letting our habits shape our lives. The deliberate act of setting norms gives us something to shoot for and a rationale for doing so.

In many ways, learning to set norms is much like learning orienteering. When we set out on a hike, we typically have a sense of where we are going. Depending on the location and terrain, we might interact with maps or markers and may need knowledge about the moon, the sun, how to use a compass, and more. Most importantly, an understanding of true north almost always serves as our starting point, as all directional needs and instructions can come from knowing how to determine true north and orient from it. The same is true of the orienteering that occurs in our internal and external worlds.

When we become intentionally aware of what we value or what is healthy, of what is most deeply important to us, we have in a sense found our own deeply personal true north. That true north rarely has to do with achievements or successes. It goes much deeper than these, although we can achieve things and experience successes along the way to setting norms that are in keeping with our most important values. Our true north also determines the kinds of norms we need to set and maintain to be our truest selves.

When we are open to a wholehearted exploration of the calling or draw toward our true north—including information gathering, the testing of our ideas, and the discernment process with self and others—we are often able to ascertain the health and authenticity of this "north." Our true north can then become the orienting point for the development of norms that will move us ever closer to a life wherein we feel a deep contentment and movement toward our personal sense of greatest health and pursuit. My misguided habit told me that As were the sign of success and smartness. My true north norms, however, are rooted in the value I place in being a person who is open to new ideas and experiences and who welcomes people with radical hospitality. My true north has nothing to do with grades—but this can be easy to forget.

Athletes, scholars, and entrepreneurs regularly use norms to help them move toward goals. Every parent who has ever read a parenting book or human who has engaged a self-help philosophy is acquainted with them as well. The primary characteristic of norm setting as distinct from habit formation is the intentional selection of what I like to think of as "spotting points," which serve as tangible manifestations of the true north toward which we desire to move. Dancers use spotting points to maintain their balance and hold specific physical spaces while executing choreography by choosing a stationary object or spot out in the distance upon which to fix their gaze throughout the dance. When they spin, they seemingly "snap" their heads to maintain a focus on their spotting point. When they move about the stage, they do the same. In so doing they avoid dizziness and maintain the expected formation of the dance. Many other athletic disciplines use this technique. Musicians use spotting points of sorts when they tune their instruments to a single note.

Spotting points don't just help orient us; they also offer us directional guidance. Where habits lead us based on patterns we've established, often by happenstance, norms provide beacons of where we'd like to go. This is the difference between doing something because we've always done it and doing something because we've chosen to do it. Spotting to behaviors or aspirations that will enhance our health and functioning provides us an opportunity to think creatively. If we want to move toward something and have a vision of what that is, we must come up with ways of accomplishing this. If we are spotting toward greater physical health, we might determine small actions we can take to move us in this direction. Perhaps we will let go of certain things we eat or ingest and choose healthier fare. We might begin moving our bodies more or reduce sources of stress that take a physical toll. Even as these smaller actions change and morph, we are constantly using the vision of greater health as a spotting point to keep us on track.

Whereas the distinction I am suggesting between habits and norms is somewhat arbitrary, it can be useful to the way we think about how we spend our time and other resources. The habit of rolling over when our alarms go off, grabbing our phones, and scrolling through everything that has accumulated there overnight is automatic and easy. If, however, we begin to find that this habit is taking more time than we would prefer, is shaping the trajectory of our day for us, and is leaving us less available to the beings with us and to our surroundings, it might be worth setting a few norms to help shift our behavior. In this scenario, we might set the norm of no phones in bed, or engagement with devices happens only after a few moments of peace and quiet. Habits require nothing to keep them going. Norms, however, force a sense of planning and preparing. In this example, norm setting might be driven by a desire to begin each day with a greater sense of calm and might require the acquisition of an alarm clock, the preparation of certain mindfulness tools to engage in place of technology, and more.

Ask individuals who struggle with a habit that has the potential of hurting them if they would rather have never started the habit or try to break it. Most will tell you they wish they had never allowed the habit to form. Norms are the tools by which we can keep harmful habits at bay and break those habits when we realize they do not serve us well. The difficulty is that norms are far from easy to create and maintain, which brings us back to the realities that began this chapter: It is easy to establish habits. It is easier to establish healthy norms than to break unhealthy habits. It is not easy to establish healthy norms.

We all live according to hundreds of intentionally chosen and evolved habits and norms every single day. Some of them move us toward health and wholeness, whereas others undercut both. Many of them are at work outside our conscious awareness, making them very difficult to change. Once a habit is firmly in place, many other smaller habits build up around it. These create a web of complexity that makes habit-breaking much more complicated than we typically realize. For instance, a smoking habit involves both physiological dependence and effect; that same smoking habit also includes social and behavioral effects that the smoker finds rewarding. Going outside for a smoking break offers camaraderie with other smokers and affords access to fresh air. In addition, the deep breathing involved in inhaling while smoking is a powerful secondary reward. Giving up the habit thus means giving up some elements of satisfaction and routine. In fact, the layered and complex nature of most habits makes them difficult to break on nearly every front. While norms can be tricky to define and difficult to establish, I believe they are easier to form than habits are to break!

There is particular relevance in this idea when applied to our habits regarding technology use. Today, everything around us encourages greater dependency on and engagement with our devices. Even when we intend to set clear and healthy parameters for our use, the convenience factor as well as the cultural tide often enable us to overlook the way we burst our own boundaries. This would not be a problem except that our habits, the behaviors we repeat many times a day/week/month/year, have an impact. They shape our lives. Just as a tree that is planted within proximity of certain wind patterns is shaped by the constant bending of forced air passing over it, so our lives and selves are shaped by the prevailing winds of technology. Our avoidance of face-to-face communication, the constant flow of self-centric content that subtly confirms our preferences and biases, and our avoidance of boredom costs us our "shape." We live with a level of convenience that impacts our ability to tolerate discomfort and delay gratification. These truths and others are altering the habits by which the human brain, the sense of self, and the relatedness to others are developed.

Not all the changes are unhealthy or negative. On the contrary, there are many positive impacts that result from certain forms of technology use, even habituated ones. The consideration here, however, is how our lives are shaped by the patterns that govern our daily lives. Evolved habits around digital engagement run the risk of becoming harmful, just as evolved habits around certain types of food can cause harm. When a person's technology use is a source of intrapersonal consideration, is governed by intentionally chosen norms and boundaries, and does not prevent the person from maintaining healthy embodied relationships with the self or other, it is likely that negative impacts will be minimal. When, however, a person's technology use simply evolves, is outside of one's conscious awareness, or significantly colors the way he or she interacts with the self, others, and the world exclusively, the impact can be huge.

Unless a person has an iron-strong will, an incredibly sturdy sense of self, and an ability to be persistent about healthy norm maintenance, the shiny, hyperstimulating, constantly moving world of digital engagement is perfect for pulling us off our true north and creating habits that distract, detract, and diminish our volitional way of being in the world. Because technology is here to stay and our engagement with it is almost always a must, the establishment of norms that enable healthy living, a balance of experiences and relationships in both our digital and embodied spaces, and the development of skills related to focus, delay of gratification, and regulation are key. The next chapters of this book will offer suggestions for how, little by little, day by day, we can hone those skills, regain control of our time, and explore our own hearts to find and follow our personal true north.

10

Assessing Our Technology Use and Making Adjustments

I recently attended the screening of a film about technology. It was produced by a highly regarded filmmaker and included many of the topics we've explored in this book. A few years back, I sat on a panel following a similar movie screened at the same theater. While the credits rolled after each of these films, I looked around to see who was in the audience with me. Not surprisingly, we all looked a lot alike; and as I scanned the faces of those present, I couldn't help but think, "So, now what?"

We had all come together out of a shared interest in what this director had to say about how the internet was created and how it is now shaping our experiences in the world. We had listened to researchers and specialists, parents and designers, academicians and scientists, and we'd been fed a lot of data. But we were offered nothing new in the way of a challenge to become proactive in response to the information. What struck me was that this is often how I feel when attending technology events.

Given that I dedicate a huge amount of time and energy to reading, thinking, listening, and talking about everything related to the presence of technology in the daily lives of humans, I feel passionate about empowering people to become active and intentional about how they engage with their devices and the digital worlds those devices deliver. While I understand the desire of many to either adopt or resist everything that comes with engagement in the digital domain, I can no longer support either approach. As I've said before, technology is here to stay, and it is neither all good nor all bad. It can make our lives better, and it can make them more difficult. Its impacts are real and complex, both in the positive and negative domains. I believe it's time to stop claiming a space on the "pro" or "con" team and get active with what we know. It's time to take a good, long, shame-free look at our own engagement (as well as that of anyone we are responsible for helping shepherd into adulthood) and to make some conscious decisions about what place we want technology to hold in our lives and the lives of our communities.

A BOX OF *LIFE* MAGAZINES, CONVENIENCE FOODS,
AND THE TECHNOLOGY PYRAMID

Several years back, when I first began consciously observing and digging into the research around how technology use might disrupt relational and personal depth and maturity, I became very depressed very quickly. The news didn't seem good. To improve my outlook and mood, my kids encouraged a technology-free evening. We got out that box of old *Life* magazines I mentioned earlier in the book in the hopes of thinking about simpler times. As referenced earlier, the ads claiming that powdered orange drink was more nutritious than orange juice, foods fried in vegetable oil were more healthful, and that is was never too early to start your babies on all manner of sugary drinks told the story of American's tendency to latch on to trends affording convenience. Not a country of moderation, we Americans bought into false claims and repeatedly turned what could have been a fine side dish into the main course meal. Ten to twenty years later, scientific research caught up with the public health crisis caused, at least in part, by the overconsumption of convenience foods. Might a similar dynamic be at play with digital devices? It seems rational to wonder how our full-scale adaptation to technology-based education, entertainment, and communication might be creating a reality ripe for negative consequences in ten to twenty years that will be difficult, if not impossible, to curb.

To help Americans make healthier food choices and counteract the widespread epidemic of obesity and other health concerns related to convenience foods, the US government created the food pyramid. While this tool hasn't solved all of our problems, it has gone a long way in helping us consider what a healthy diet might look like. Similarly, I suggest that the first step of any plan to alter our technology engagement begin with a consideration of what I refer to as a "technology pyramid."

Just like the older version of the food pyramid, the technology pyramid I suggest recommends a greater proportion of healthy engagement with platforms, devices, and apps that deliver technologies that have less potential to hurt or limit us—and less engagement with those that have greater potential to hurt or limit us. (You can find a sample "Technology Pyramid Worksheet" and a link to a downloadable form in the appendix.)

When exploring the idea of a technology assessment using the technology pyramid or any other possible tool, I stress considering the quality of the devices, platforms, and apps being engaged. Ideally, more time should be spent in digital spaces that encourage connection to embodied people within one's social and relational sphere or ones that are truly educational in nature. The educational sites, apps, and programs that are beneficial to spend more time with are those that best reinforce effort

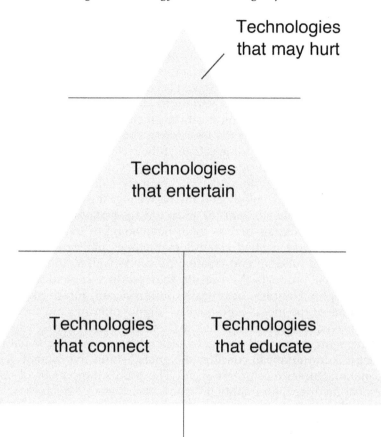

and allow mistakes to be a part of the learning and creation process, thus reinforcing a growth mind-set. With the technology pyramid model, I encourage spending less time with purely entertainment-based platforms than with connecting and teaching platforms. If someone is going to use digital devices for entertainment, my recommendation is to choose sites and devices that allow for the cultivation of focus and delay skills as well as the encouragement of creativity, effort, and mistake making within the context of healthy and pro-social environments as part of learning acquisition. When the chosen entertainment sites and devices center on exposure to violence or inappropriate, sexually explicit, and commercially driven content, real effort should be made to minimize engagement and to counter the impact by embodied experience. This is especially true for younger users.

The importance of the *quality* of the digital content we interact with cannot be understated. Just as with the food pyramid, the quality of what we ingest is equally as important as the content of what we ingest. Fresh fruits and vegetables provide the most potent vitamins, minerals, and nutrition. Flash-frozen foods are next best, followed by sodium-drenched canned vegetables; deep-fried variations are at the low end of nutritional value. In other words, eating a can of green beans is better than eating a bag of potato chips, but it's less potent than eating a bowl of freshly picked green beans. The same is true of the technology we ingest. Choosing high-quality, well-produced media in every domain of the technology pyramid will pay off with a fuller, richer, more balanced life.

Evaluating the technology and media we ingest can be difficult. Video games and porn are some of the most engaged platforms at present. As a result, the industries tied to them have some of the deepest pockets in business. These content creators (sometimes platform creators) have access to huge funding and creative power as well as to users ready to invest in their products. As a result, they produce content and products that are highly complex, beautifully designed, and fun to use. They are likely some of the most "high-quality" digital offerings out there. This is not, however, the type of high quality to which I am referring when I suggest engaging more well-produced technology and media. What I mean is that it is important to consider the goals behind the content and how the manufacturers reach these goals. The production values of my own personal favorite high-quality TV show, *Mr. Roger's Neighborhood*, might be considered laughable by most content creators and huge swaths of the general population today. The content, however, is excellent, presented in intentionally chosen ways at reasonable speeds, affording the creative interaction of the viewer.

So, how does one go about evaluating any digital media or platform? It can help to ask the following questions:

- What is the creator's goal?
- Does engagement with this space teach me something?
- What does it teach me?
- Is this a value, skill, or informational domain that improves my life?
- If this is purely frivolous fun, are there any negative impacts I should be aware of as I engage with it?
- Do the creators of this content benefit directly and monetarily from my repeated or prolonged engagement with it? Am I encouraged to make frequent in-app purchases? Am I required to pay for frequent upgrades?
- Am I required to watch ads repeatedly during engagement? If so, what are the ads for?

- How does the flow of offerings to me in other digital places change after I've interacted with this platform? Basically we're asking, "How does my engagement here in this space impact the algorithm that determines what is offered to me online?"

A good rule of thumb is to take a look "underneath" the sites and apps being interacted with. For example, many art museums host beautifully designed websites that contain all kinds of educational information. Rather than doing a general search for design or art information, you might find these museum sites to be more reliable, more deeply considered, and more well designed than wherever you happen to land after a general search. Sometimes paying for access will mean less exposure to mindless ads or never-ending streams of content intended to create a physiological desire for more. Nowhere is this truer than with games. Game conglomeration sites offer hour after hour of frivolous and highly stimulating games with ads that must be watched or clicked on in between. These sites definitely consider the user the *product* and the advertiser the *client* and hope to keep people on their site longer, regardless of content quality. Similarly, video-sharing sites such as YouTube, where more than four hundred hours of new video are uploaded every minute, are rife with content meant to excite and hold our attention hostage. Turning to videos made with greater intention, goals, and production quality will train our minds to expect and desire content that is not merely attention grabbing.

That being said, video sites such as YouTube and Vimeo can be amazing places to go to learn a new skill. In my own research, people have learned how to cup stack, spackle, knit, crochet, fix a microwave, repair cars, do makeup and hair, create art, budget, cook, craft, draw, put on tire chains, fix a smartphone, dance, frame a door, repair plumbing, record music in a home studio, sew a princess cape, and beatbox via online tutorials. What a fantastic example of using an online resource to create greater embodiment in life!

TAKE ACTION

To get your creative juices flowing, here are some other things to learn via online tutorials:

- Origami
- Portrait drawing

- Paper airplane folding
- Clothing design

- Putting together wardrobes
- How to build a go-kart
- How to juggle
- Double-dutch jump roping
- Mastering a front roll or cartwheel
- Brewing root beer or kombucha

- Understanding the stock market
- How to play kendama
- How to make tin-can stilts
- Slime making
- Cooking the perfect egg
- A million other creative ideas!

Another way of increasing quality of experience is to increase the quality of your own investment by creating content instead of just consuming it. We must admit, consuming within digital spaces is largely passive. We watch videos; we surf and scan sites. We merely take it all in, or absorb content, which typically requires little engagement of the mind or body. On the other hand, when we create, our minds and bodies are engaged! Thus, creating with technology and within digital spaces has the potential of improving the quality of our experience. Learning to code, creating websites, mastering the process of making music or videos online, and writing a blog are all examples of being creative in ways that could also enhance the quality of our interactions with technology and the internet.

The only real way to be assured of interaction with high-quality content is to do some research. Websites such as CommonSenseMedia.org go a long way toward providing nonbiased reviews from both children's and parents' points of view. Often, a quick web search of the platform, hardware, software, or apps with the terms "reviews," "scholarly research," or "potential risks" can uncover the backstory or potential pitfalls of said technologies. Asking others is also a great idea, but parents especially must remember that not all families make the same choices for their engagement. Ultimately, it's up to each of us to understand and make informed choices about the quality of what we interact with.

But embracing a balanced, moderate stance regarding technology is not easy. Working toward a life of engagement in the digital domain without full immersion—without filling up 100 percent of the pyramid of our day with digital activities—actually requires discipline and critical thinking. It's easier to live a reactive life, allowing information, entertainment, marketing efforts, or news reports to dictate our feelings and responses and eat up all our time. Speaking of time, it's time we begin to moderate our full-fledged adaptation to our devices!

ANOTHER TECHNOLOGY ASSESSMENT TOOL

One of the most common questions I receive at talks or via email is "How do I know if my child/spouse/partner/self is overusing technology?" In other words, "How much time is too much time to spend with screens?" Over the course of the dozen-plus years I've done this work, I have struggled to come up with a concise answer that has its roots in research and reason rather than in opinion and sensationalism. I wish the remedy for our overuse were as simple as limiting, or controlling, our time. Instead, many factors go into determining how our technology use is impacting us and how we might become more intentional to prevent negative effects.

After playing with all kinds of complex assessment possibilities and researching the various benefits and pitfalls of digital engagement, I created a simple tool for determining one's technology engagement and the way it may be impacting a person's overall life. This tool allows us to gather qualitative data about the way we are engaging with technology in a way that helps us moderate use. By taking some time and energy to become aware of the areas in which our use is moderate and healthy—and where it is not—we are able to find ways of adjusting our digital engagement, providing counterbalances for our tech time and expanding the richness and relevance of our embodied lives.

The tool can be helpful to everyone. I've seen it used with great impact by parents, university students, psychologists seeing clients who are struggling with digital addictions, pediatricians trying to determine if too much screen time is affecting the health of their young patients, and more. Who couldn't use a healthy dose of honesty about our habits? And everyone's digital dilemmas are different! The only way to deal effectively with those dilemmas is to accurately understand them—and that involves starting with an honest assessment. Once we understand the nature of a particular problem, we can determine a particular solution.

The five components of this assessment tool are easy to remember, as I have tied them to the ABCs, with a twist (T) at the end. Using your hand and beginning with your thumb, say one letter for each finger: "A. B. C. D. T." Each letter is a prompt for the five categories that can be issues when it comes to dependence on technology and will be described in a few pages:

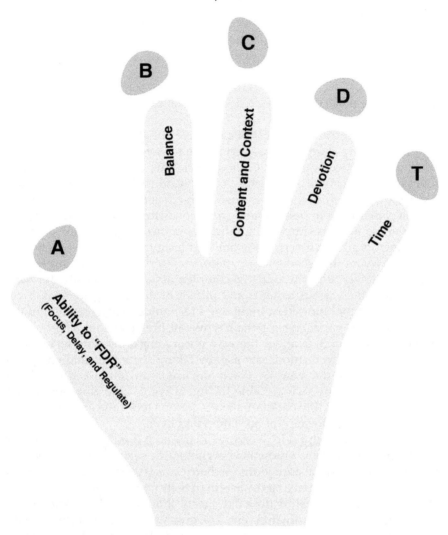

By writing these letters down the side of a piece of paper and making columns for notes, we can create a "Technology Assessment Chart" and get a general overview of how we're doing. Here's a sample:

| Name / Date: | doreen 1.24.2018 |

A	**Ability to Focus**	3 – Finding myself distracted by email while working.
	Ability to Delay	4 – Finding myself tending to things immediately when they could wait.
	Ability to Regulate	2 – This should get better as I become more intentional about my focus and delay. Keep tabs on this. Meditate!
B		Balance	3 – Given the requirements of online work right now, I'm not being intentional about in-person get-togethers.
C		Content / Context	2 – I need to find some ways to lighten up more. I need lighter-hearted content more of the time. What broad category of technology do you gravitate toward? Strategy, Logic, Social/Relational Media/Pop Culture, Creative Arts, Body/Kinesthetic
D		Devotion	1 – I feel good about this area. I am devoted to experiencing things without reviewing them to death!
T		Time	5 – Writing this book has meant that I am with my computer far too much. I need a lot of work here!

This chart will look different for everyone. Some people using it might want to make a number scale to rate and record how they are doing in each area. For example, in the "A" row, they might rank their ability to focus as a 4 on a scale of 1 to 5, where 1 means "low ability" and 5 means "high ability." Their delay skills may be self-assessed as a 4; their regulation skills may be a 2. Moving to the next row, they may find that their attachment balance needs more work, scoring a 4 where 1 means, "My relationships are balanced with a similar or greater number being nurtured in embodied spaces compared to those in digital spaces." A rating of 4 might result from the realization that most of this person's connections are maintained online rather than in physical spaces. A number-rating system might not work for some. Instead those individuals might make notes in each column, using words and descriptors to understand the tech engagement and impact.

I have found that doing this assessment on a monthly basis can be helpful as my clients and friends work to make their tech use more moderate and healthy. To remember to complete it, some even ask to receive this simple assessment chart monthly via email. You can find one to copy and use in the appendix.

MAKING ADJUSTMENTS AFTER ASSESSMENT

Assessments are only helpful if we become active with them. Becoming active with them however, is downright difficult. In fact, the change-making part of taking stock is often actually painful. We can identify the things we want and need to change, but determining where and how to start is frequently overwhelming in every way, and the costs feel immense. The reality is, though, that making changes does not have to involve a total time or energy transformation. Neither does it need to be overly costly. The goal should be to make small, meaningful, and sustainable changes over time with readily available resources. The more targeted we are in addressing the specific areas where we need to make changes, the more specific our efforts will be. This means that the more detailed we can be about where our tech engagement might be negatively impacting us, the more likely we will be able to regulate the problem, thus making increased space for a fulfilling embodied life. To make this happen, the effort will need to be real and committed.

The rest of this chapter is designed to help with moderating technology use in order to develop our abilities to focus, delay, and regulate; achieve attachment balance; moderate our devotion to our devices and the content they deliver; and effectively manage our time. To that end, the following pages provide a basic set of ideas for objects to engage, activities to try, and questions to ponder relevant to specific "need" areas. The ideas, activities, and questions are intended to get us thinking about creative ways of making our embodied life rich in ways that will help our technology use stay moderate and centered on engagement. Of course these suggestions are jumping-off points. Our own creativity or the creativity of those we know will provide the best solutions for how to make our technology use healthy and supportive of a well-lived, fiery, rich life.

ASSESSING AND ADDRESSING "A": ABILITY TO "FDR" (FOCUS, DELAY, AND REGULATE)

As discussed earlier, being able to focus one's attention, delay, and self-regulate are critical to optimal human development.

Building the Ability to Focus

Recall that "focus" refers to the ability to hold sustained attention toward one object (or a small field of them). These objects can be tangible or can be internal constructs such as thoughts, ideas, or feelings. In essence, focus is built on a developed capacity to tune out distractions. In other words, focused individuals can say "no" or ignore input and stimuli that compete with the object of their attention. Whereas some temperaments seem to demonstrate a natural proclivity toward attuned focus, others need more practice.

More than ever, people's capabilities in this arena are challenged. Many individuals with whom I speak discuss distractibility as a primary frustration in both themselves and others. Particularly powerful at impairing our focusing skills are those technologies that reward us for scanning the environment and having a wide range of attendance to the "field"—as opposed to those technologies that encourage focus on one task, character, or directive. Much of the digital domain is built specifically on the promise of helping us task switch with great efficiency. This is in direct opposition to focus.

When deficits in the domain of focus reveal themselves, it is important to take intentional steps to develop this ability. Being able to ignore or tune out some stimuli to focus on other sources of input is central to satisfaction and success in relational, academic, vocational, and spiritual pursuits. The goal in acquiring focus skills is to build an ability to direct attention toward a specific and limited area and sustain the focus for an ever-growing period of time. The kinds of questions that get at measuring this ability sound like this:

- Can I complete a given task without diverting my attention from it?
- Can I know that a message has come in yet not check it until I have completed what I was focusing on when it arrived?
- Can I maintain a conversation with someone without referencing my device for at least thirty minutes?

To develop focus skills, we must create opportunities to do one thing at a time, turning away from stimuli that compete for our attention. While daily life used to offer such opportunities as a function of the sheer opportunities and resources available to any one individual, we no longer live in a world that offers us swaths of time for boredom and generating our own fields of focus. For this reason, if assessment uncovered a need for greater focus, it is important to build in daily opportunities for focused attention and deep work across a number of domains.

A task that many call "redirection" is important for building and sustaining our ability to focus. *Redirection* refers to the ability to notice our distractibility, name it, then return our focus to the task (or idea, etc.) at hand. Perhaps you are working on writing an important email. While you are composing, several texts and emails arrive. Redirection skills would allow you to notice that the new messages are distracting you, name this truth, determine that waiting to attend to them is a viable option, and return your attention to the original email. Your internal dialogue might sound like this: "It's really hard to stay focused. I want to read the new messages. I need, however, to stick with this until I'm done, and doing so will be rewarding."

With devices that alert us to all activity and minds that drift away from objects of focus and onto distractions in our environments, we need a well-developed and practiced ability to redirect our attention back to the task, thought, or experience at hand. Simply deciding that we will have better focus and intending to stay on task will never be enough. These skills are supported by neurological wiring. If they aren't practiced, the wiring will not be robust enough. If we do not actively protect time and create experiences with which to practice, focus and an ability to redirect will always escape us.

My friends Lynea and Jim Gillen are the founders of Yoga Calm. In keeping with the mission of this world-renowned yoga program for children and adults, they created an innovative class during which children can learn focus through creative play. They call these "Jedi" classes and use the concepts of Jedi training, illuminated in the *Star Wars* empire, along with toy light sabers to teach young, easily and highly distracted children how to stop, pause, and focus. With arms at their sides and light sabers in their dominant hand, Lynea and Jim teach kids belly breathing as they stretch their arm and light saber out to the side and up. At the top of the arch they exhale, bringing the sword to their chest. After many repetitions, children learn to do that activity like a "Real Jedi" in their minds. Needing no prop, they eventually can redirect their attention from distractions to the present moment, breathing deeply and focusing on what is at hand. We all need this type of practice!

When we train ourselves to meet distraction with a deep "Jedi" breath, we are giving ourselves the space to direct our attention to a chosen spotting point. On the inhale we can see and name the distraction before us; on the exhale we can blow it away from us and return our focus to the task at hand. If we practice this and can trust that anything of importance will come back to us when it is time, we can redirect our attention to the task we were working at prior to the distraction. In the early days of our practice, we may find ourselves naming and blowing away our distractions more than finding a settled focus. Do not fear: this practice will eventually pay off as we learn to notice and name our distractions and redirect our attention.

TAKE ACTION

Here are a few activities that encourage FOCUS:

- Mindfulness meditation
- Yoga (including "Jedi training")
- Listening tasks/challenges
- Plays/live theater
- Solo imaginative play
- Puzzles

- Contemplative prayer
- Reading a paper book

- Classical music concerts
- Chess and strategy games
- Memorization tasks
- Balance games and tasks

Building Delay Skills

Like being able to focus one's attention, having the ability to delay is developmentally optimal for humans. The ability to *delay* refers to our capacity to pause and wait, especially when coupled with a desire for a particular experience, sensation, object, or outcome. When we can master this skill, we realize we can face the discomfort that accompanies the need to wait, knowing that it doesn't preclude our (eventually) getting what we want or need. This is true if those desires refer to physical possessions, relational wishes or needs, or actions or behaviors.

Similar to how our twenty-four-hour-stimulation-driven culture impairs focus, an individual's competence in delaying gratification is deeply impacted by having most things readily available at all hours of the day and night. Thanks to incredibly powerful technology, we can purchase whatever we want whenever we want it from nearly anywhere, we are able to communicate with a friend across the world with no lag time, and we can learn the answer to any puzzle with a quick search. With so few reasons to delay and work through things, people simply don't do it—and we get more and more out of practice every day.

In previous centuries, people developed the skills of focus and delay as a function of living in a world where they were forced to wait. Through the process of waiting, alongside the presence of fewer distractions, people simply learned to focus on things inside their own minds or within their immediate surroundings. Because opportunities to be constantly entertained or occupied did not exist, this dearth of stimuli created a natural lab for boredom tolerance and sharpened attention—at least some of the time. Subsequently, the requirement that one must wait for either external resources or internal insight led us to value the resource of insight more deeply. When something is difficult to attain, it is often more highly esteemed.

Any activity that allows an individual to pause between stimulus and response will build delay capabilities. The ability to delay is important because it provides us with margins within which we can unfold discoveries or make intentional decisions. In many ways, what I mean is that an ability to delay might be just the antidote to impulsivity and shallow relationships. When we can delay, we engage our consciousness before acting, and we have a greater ability to know our own selves and to encounter the authentic selves of others.

While delay of gratification specifically gets plenty of press, delay as a general skill is related to many other human experiences. Delay, in all domains, helps us gain a sense of sturdiness that serves us as we think, feel, and act. It also helps us esteem and value the efforts we make and the answers or solutions we find.

The important thing with learning delay is to remember that a little practice, done consistently, can go a long way. It's not important that we do these suggested activities all the time, but we must do them consistently some of the time. Even doing research once via a phone call or an in-person library visit benefits us doubly by (1) increasing our skills in delaying while we wait to acquire knowledge and (2) making us aware of and grateful for the speed with which we gather data every day.

TAKE ACTION

The following activities teach DELAY skills:

- Making a phone call, leaving a voicemail, and waiting for a response
- Watching a television series in weekly episodes rather than binge-watching it all at once
- Waiting in line, doing the shopping, completing a drive from one location to another, or eating a meal—without interacting with a phone, laptop, television, or electronic tablet
- Waiting a preset amount of time between the urge to impulse-buy an object and actually purchasing it

- Writing back and forth through "regular" mail with a pen pal or friend
- Shopping in-person rather than online
- Forcing a waiting period between screen times (e.g., thirty minutes with screens followed by a mandatory thirty minutes without before screens can be engaged again)

The following activities help children develop DELAY skills:

- **Play stop-and-go games** such as Red Light/Green Light, Musical Chairs, or Duck, Duck, Goose.
- **Narrate waiting time.** For example, when a child wants an object or activity, create a period of waiting beforehand and narrate this to the child. For example, say, "I know you'd like to play with your Legos right now, but we also need to finish our meal. Let's see how it feels to wait. I'll help you do this!"
- **Use a timer or clock and explain the concept of waiting.** When it's time to stop waiting, really make note of how good it feels to have waited and then to get to move on to the desired activity.
- **Wait as a family.** Choose some desired activity or purchase and make a plan for how you will, as a family, save time or money to invest in it. Mark your progress with a chart or on a calendar; really celebrate when the waiting is accomplished and the delay ends.

Learning to Self-Regulate

As noted earlier, the ability to *regulate* refers to the ability to know and consciously engage our internal emotional and cognitive states. Interestingly, this skill is directly related to our ability to focus and delay. People who can regulate are able to understand the interplay of the external world and their internal responses to it. In fact, our regulatory skills allow us to observe the ways in which our thoughts, feelings, or experiences are evoking responses in us and then to address those thoughts, feelings, or experiences with action or physical self-soothing.

Impulsivity is the hobgoblin of regulation. When we have developed patterns of impulsivity, we are prone to move *from* where we are rather than *toward* something better. Unless it has been actively employed in the service of pausing, impulsivity screams, "Do whatever is necessary to change what I am experiencing now!" Regulation, however, adopts a more reasoned stance. A person who can regulate mindfully asks:

- What is happening in or around me?
- Why am I reacting the way I am?
- What can I do to get through my thoughts and feelings?
- How can I find comfort or resolution within myself rather than outside myself?

For example, when someone faces disappointment but can make sense of that disappointment cognitively, offering comfort and understanding to herself, she is demonstrating self-regulation skills. When someone encounters failure or hardship and knows how to think and work through the difficulty without its having undue impact, that person can regulate. When a person feels strong feelings and can assess them with an eye toward resolving them, regulation has been mastered.

Leah Kuypers has developed a fantastic curriculum that addresses helping children regulate within the classroom. Kuypers describes four color zones that coincide with different feeling states. The red zone is characterized by heightened states of alertness and intense emotions. In the red zone, emotions such as elation, anger, rage, fear, and terror are experienced. When people are in the yellow zone, they are still in a heightened sense of alertness with strong emotions, but they feel a greater sense of control and capability. Feelings such as stress, excitement, agitation, and anxiety fit here. The experience of being in the green zone is that of being calm and alert. The best learning and living happens here, as people feel content and ready to face their present circumstances. The blue zone corresponds with feelings of sadness, sickness, tiredness, and boredom. The helpful handholds of colors and conceptualization of zones give us words for how we are feeling and enable us to brainstorm active ways of moving ourselves from one zone to another. When this can happen in conscious ways within a healthy self, we can get ourselves to better regulated states with relative ease. When we have a shared vocabulary such as this, our family constellations and other relational systems live in better harmony, as we can understand and communicate about the various zones we all move between.

Unfortunately, however, as increasing amounts of our lives are lived outside ourselves, we become less and less likely to develop or rely on self-regulation skills. Not only do we simply ignore ourselves and our current realities by spending inordinate amounts of time staring at our screens, but we also look to the size of our followings, the number of likes we acquire, and the range of our digital reach to decide how to feel about ourselves and the world around us. As psychiatrist and author Victoria Dunckley so aptly describes in her article "Electronic Screen Syndrome: An Unrecognized Disorder" in *Psychology Today*, we have come to depend on screens rather than develop our own capabilities. Above all, rather than developing an ability to regulate, we distract ourselves with near-constant stimulation.

On the contrary, when we can regulate ourselves, we can moderate and traverse the intersection between our internal experiences, our physiological realities, and the external world of expectations and stimulation. Regulation includes an ability to read the environment both within and outside ourselves in such a way that we can act in manners that honor the self and respect the setting that the self exists within.

The ability to self-soothe may be the single most important skill for the development of self-regulation. Self-soothing is built on the abilities we've just discussed and can only occur when an individual can pause in response to an emotion, thought, or impulse to focus on what is occurring in the present moment—and can move from one emotion zone to another. There are self-soothing behaviors that are healthy and those that are not. Healthy soothing behaviors bring us to places where we can function in integrated and regulated ways that lead us to resolve any present conflicts or areas of unrest and move forward. Time-outs, intentional breathing exercises, actions that enable us to "clear our heads," physical activity, and more are examples of healthy self-soothing behaviors. Self-soothing behaviors such as substance use, restriction of food, and self-injury are examples of self-soothing behaviors that, while seeming effective in the moment, can cause harm.

As with opportunities to develop focus and delay, regulation skills are best developed with intention. When we need the ability to regulate that is *not* the time to try to develop the skill to do so! I highly recommend that we find ways of building and practicing self-regulation skills in times when we are not dysregulated. By doing so, we build self-soothing muscles we can employ when we really need them.

TAKE ACTION

The following activities teach SELF-REGULATION skills:
- Deep-breathing exercises
- Progressive relaxation
- Mindfulness meditation
- Yoga
- Psychotherapy
- Spiritual direction
- Contemplative prayer
- Self-help or human development books

- Self-discovery classes
- Use of prayer beads, labyrinths, and mandalas
- Coloring/sketching
- Physical exertion in the form of exercise that is pleasurable and releases tension
- Some forms of personality tests (Meyers-Briggs, Enneagram, etc.)

ASSESSING AND ADDRESSING "B": ATTACHMENT BALANCE

Attachment balance refers to the proportion of a person's connections to others in embodied life compared to their connections in digital spaces. If our primary attachment to others is through digital means, it is important to note whether the communication happens person to person or through characters built online. If it is mostly mediated by characters (e.g., avatars) or very carefully curated profiles built online, it is important to take a look at how our wishes, disappointments, and general beliefs about ourselves are reflected or compensated for in the characters we have created. If these representations of our selves are far from authentic, we might appropriately wonder if the connections made to them feel secure or if an ambient awareness of "being found out" might impede the relationships.

A primary feature of attachment balance is the maintenance of communication skills necessary for embodied person–to–embodied person communication (i.e., conversational skills, eye contact, active-listening skills, verbal articulation, and comfort with pauses). A lack or deficit in these skills does not necessarily mean a person is overly attached digitally. It does, however, indicate that an intentional increase in embodied person-to-person contact and skill building that this kind of practice can achieve may be in order.

For us to achieve balance in our primary attachments, we need to be able to accomplish four tasks:

1. Establish a community of meaningful relationships, at least some of which are engaged in embodied forums and with authentic presentations of the self.
2. Develop the ability to communicate effectively and appropriately in embodied spaces (face-to-face or, at a minimum, voice-to-voice).

3. Develop the ability to communicate effectively and appropriately in digital spaces.

4. Maintain a flexibility about the most effective way to communicate across different situations and an awareness of our preferences in communication and connection (in-person or digitally).

Today, it is easy to shift a huge bulk of the maintenance of our relational work to digital spaces. Whereas there is nothing inherently bad or wrong about this, an ability to interact with people in embodied ways is important for our long-term health. When I taught parenting classes, I always encouraged parents to make sure they provided opportunities for their children to talk with people of all ages, genders, and personality types both about something specific and about "nothing at all" (aka "shooting the breeze"). When we have active practice in engaging others and have evidence that we can live through the sometimes-awkward realities of in-person communication, we can risk embodied connection.

When these skills are rusty, we can be at greater risk of consciously or unconsciously avoiding such connection, and we can become nervous about interacting with others. As anxiety about in-person encounters increases, so do our feelings of incompetence and fear of the unknown. A fantastic therapist I know invites socially and relationally anxious individuals to fill a coin purse with change and take it into a coffee shop. After ordering a drink, making eye contact while doing so, the client is challenged to drop the change, pick it up, count it, and hand it over, thanking the cashier verbally. In this experience, the client must face up to fears about competence and skill and, in doing so, build both.

If we rigidly resist communicating digitally, we may face similar anxiety and disappointment in relationships. The reality is that many people rely on the ease and speed of texting. Individuals who prefer voice-to-voice communication in person or on the phone cannot always expect others to comply with their preferences. We must all move *toward* one another and, at the same time, be willing to flex and adjust if we hope to have a complex relational life.

To have and maintain a modicum of balance in our attachments, we need to hone and maintain our abilities to both speak and write, to meet in person, to speak on the phone, to email, and to text. We must seek out connections in our embodied spaces as well as in our digital ones and benefit by trying to spend at least a portion of our relational energy in embodied encounters.

TAKE ACTION

Here are some creative ideas for increasing ATTACHMENT BALANCE:

- Use online sources to find individuals with similar interests with whom to spend time in the physical world. Meetup.com is a great resource for this.
- Attend lectures and community talks (public libraries, bookstores, and local universities are good places to look for these). Challenge yourself to talk to at least one person while there or to ask a question in the group question-and-answer session.
- Practice making conversation with people in your natural surroundings such as grocery store clerks, baristas, waiters, librarians, and more. Be intentional about your own eye contact and nonverbal cues, and don't make assumptions about those of others.
- Make at least one phone call for every ten texts you send.
- Learn and teach digital citizenship. Learning how to communicate effectively in writing is especially important now that so much of our communication is done in this way. Texting and instant messaging can risk diminishing or altering our articulation and communication skills. Digital citizenship training can keep healthy and proper communication skills in place when communicating in digital spaces. CommonSenseMedia.org has a fantastic curriculum that can be used for this.
- Practice eye contact. I mentioned this before, but it's worth repeating. Find people in your relational circles who will practice gazing with you. Set a timer and try to silently maintain eye contact. This is hard but worth it. Every. Single. Time.
- Talk with those closest to you about why they prefer the methods of communication they do, and work diligently to understand them. At least some of the time, challenge yourself to communicate with them via their preferred method as opposed to yours.

ASSESSING AND ADDRESSING "C": CONTENT

During my talks, I raise the finger associated with "C" in my hand model (the middle finger) and comment upon how much trouble I can get into by raising this finger alone. The same is true with the *content* we consume!

While assessing this digital domain, we must consider the many different layers of quality and subject content that dominate the games, websites, and social networks with which we interact. Is the content driven by strategy, competition with self or others, intellectual pursuits, risk-taking? Do we engage with a wide variety of digital spaces, or are we preoccupied with one certain type of media or technology? Pay special attention if the content is solely focused on violence, rigid or fixed gender stereotypes, or sexualization and/or is marketing driven—all of which can tend toward the unhealthy. Referring to the technology pyramid, if most of our exposure is to low-quality, sales-driven, or violent or sexualized content, we will likely experience more negative effects than if our use is with high-quality, non-commercialized content.

At first blush, content can seem the easiest change regarding our technology use. Taken together with time, however, the content we engage massively shapes our experiences online, therefore impacting our physiological, emotional, and relational well-being. The type of content we engage actually both exposes and reflects our preferences along many domains, including entertainment, learning style, and communication. This complex reality means that we could expose ourselves to inordinate amounts of certain types of content without considering the other material that comes with it. As noted earlier, individuals who enjoy violent first-person shooter games may do so largely because of the richness of the graphics and the intensity of the strategy requirements. The potential impact that the violence may have is likely a second- or third-level consideration in the choice of platform. The same is true of any platform that employs repetition with perfectly placed moments of novelty. Who cares if the mastery we are gaining involves lining up candy? The repeated opportunity to do so in new ways keeps us coming back for more.

If humans were capable of maintaining moderation, this would not be an issue. The truth, however, is that once we find something we really like, we often tend to overdo the amount of time and energy we sink into it. We really enjoy flipping through social media to stay on top of what is going on with our community, so we end up investing a bulk of time and energy there, over time making it a cornerstone of our relational self. With gaming, we find content that suits us and submerse ourselves in it, similarly turning it from a point of entertainment into a place wherein we play, engage our strategic mind meaningfully, and connect with clans of others who do the same, leaving little time for experiences in other spaces that might make us more well-rounded individuals.

When we find that our engagement with a social media platform has come to substitute for meaningful, complex, out-of-platform thinking and acting or for embodied relational pursuits and has caused us to lose some of our skills related to both personal and interpersonal engagement, it can feel scary to give up the social media altogether. Similarly, if our minds are most actively engaged in our video game play, it's unrealistic to think we'll want to step away simply because the content can harm us. Our minds love to be stimulated, and we are often quick to overlook negative effects that come alongside positive ones. For this reason, we need to be able to tell ourselves the truth about the content in which we invest ourselves. It is crucial that we determine what the healthy parts of the content we engage are, and to also be aware of any potential negative or hurtful effects.

To do this, we must consider the "backbone" of the technology toward which we gravitate. Content can be very loosely organized into the following broad categories:

- Strategy
- Logic
- Social/relational connection
- Media and pop culture
- Creative arts
- Body/kinesthetic interests

Many video games would likely fall into the strategy and logic domains, whereas social networks would likely live within the social/relational realm. Content creation technologies may be considered media-based and artistic, whereas technologies specifically geared toward physiological experience might work by teaching a skill to be performed in embodied space or by eliciting certain physical responses, as with porn.

Because we have built our engagement with specific content into our lives, it is unrealistic to imagine simply putting it aside. In many cases our relationships with the content we engage have become support systems to our daily lives. We feel that they comfort and entertain us. Particularly if our tech engagement involves being networked with others, online social platforms are sometimes the vehicles that deliver much of our social support. Because of this, we must be specific as we work to reimagine how we might engage digital content. In addition, we must work extra carefully to put new supports in place in our embodied lives as we make any changes.

Once we can identify the primary themes driving our draw toward certain technologies and devices, we can determine what, if any, embodied experiences we might engage to balance out our tech engagement. I have included in the appendix lists that feature ideas of how to infuse embodied living with themes frequently motivating the overuse of technology as well as suggestions for how to make sure the technology engaged is of high quality and the least potentially harmful. Again, the items included in the lists are simply jumping-off points, suggestions for myriad offerings that have the potential to engage our interests in more well-rounded ways.

ASSESSING AND ADDRESSING "C": CONTEXT

As with the content we consume, the *context* within which we engage technology also matters. "Context" refers to the environment within which we engage technology. From homes (kitchens, bedrooms, bathrooms) and cars to coffee shops, theaters, stores, and more, how we engage technology within our physical contexts can either lead to an increase in pro-social behaviors or render us overstimulated and less capable of pro-social action. Referring to the technology pyramid, if we engage our devices in such a way that we constantly contextually isolate ourselves, wherein our technology pulls us out of the context of embodied living and into the world of the digital domain, we will pay a price. Granted, using technology in the context of embodied encounters can be neutral, but it also has the capacity to rob us of the ability to practice in-person encounters. That's why we must honestly assess whether we use technology to contextually isolate ourselves or to connect ourselves to others through mostly digital means. Consider that earbuds provide one level of isolation and physical locality provides another.

Questions to ask when considering the domain of context are:

- Are we consuming technology solely in isolation?
- Do we spend all our "social" time on screens, or are we able to put screens away to interact with embodied others?
- Do we use our device(s) to escape awkward, uncomfortable, or undesired contact with others who are physically present?
- Can we have an embodied encounter without referring to or accessing our devices?

Context is important because it has a bearing on two primary lifestyle outcomes. First, if we are technology dependent *across all contexts* (social and solitary, within and outside the home, while at work and at play), there is a very high chance we are forgoing important investments in social practice and conversational skill-building. These two behaviors are crucial for long-term life satisfaction and health. If we find ourselves grabbing our devices to look things up during conversations or must share the latest YouTube craze when we are with others, it may mean that we are too dependent on our devices in relation to others. Second, if we use technology solely or even primarily in isolative settings, there is a strong chance we are missing out on the input of others in helping us discern which digital spaces are and are not healthy and life-giving for us.

Making sure our home environment provides a multitude of options for creative engagement, play, and exploration will help us be able to separate from devices and engage embodied life. Being able to be bored, wait in line, be in social settings, and perform basic conversational tasks all take practice. Our current way of living does not always provide these opportunities. For that reason, we must intentionally practice separating from technology to learn to thrive in more embodied contexts. It may feel less fulfilling to verbally describe something than to pull up photos or videos to illuminate our stories, but doing so can be a powerful action toward growth. There are hundreds of other examples of similarly small risks that could reap large inter- and intrapersonal rewards.

Having spaces within the home where technology is used and spaces where it is not used can be helpful, as can setting some minimal guidelines for ourselves regarding embodied versus digital connections. For interventions regarding space use in the house, all family members must agree to participate. One simple, albeit difficult, contextual intervention is to have family members use alarm clocks, docking their phones together and away from all bedrooms at night. Another idea is to have the router switch off at a certain time at night so that all devices are internet-disabled after that time.

Similarly, the establishment of some simple but stretching norms around how one does or does not engage technology in times of solitude or connection can be helpful. Deciding on an hour per day when all devices are turned completely off is a good place to start. Choosing to leave your phone in the car or a backpack when meeting with others (as opposed to having it out and on the table) would also be a healthy place to begin.

ASSESSING AND ADDRESSING "D": DEVOTION TO TECHNOLOGY

In the West, a ring on the second to the last finger of the left hand is often a sign of marriage or commitment. For that reason, in our hand model this finger is associated with the level to which we are devoted to our devices and the digital spaces they offer us. *Devotion* in this context involves our level of complete dependence on our devices. Important here is determining the types of day-to-day living skills we have completely given over to technology versus the actions we still perform without technological assistance. The skill of getting directions and finding our way to a location previously unvisited is an easy place to begin. For example, can we find our way to a new place without a GPS or smartphone? Can we determine a route without turn-by-turn directions from Siri? Can we find an address without Google? Similarly, if we can't place an order online, can we find a way to buy what we want in a brick-and-mortar store? Can we accomplish basic functions such as waking up without a digital device? Are we willing to choose a restaurant without reading reviews or watch a movie without checking its "Rotten Tomato" score? Can we determine the weather with a quick step outside or a peak out a window rather than opening up a weather app? Are we able to be spontaneous or resourceful without our devices in any part of our lives?

If assessment uncovers that we are unduly devoted to living all parts of our life online or doing all daily living tasks with a device in hand (shopping, ordering meals, meeting people, maintaining relationships, finding places, time telling and awaking, etc.), important cognitive and physical functions may be at risk of diminishing due to lack of use. This demands that, as an intervention, we seek non–digitally based ways of living and of completing tasks—and enhanced opportunities for embodied living at least some of the time.

Some ideas for actions to help fight an overly dependent devotion to technology by engaging embodied experience include:

- Look up directions with a digital or paper map before heading toward your location. Either make notes about the directions or memorize them. Drive without being dependent on the verbal cues of the GPS.
- Try a restaurant without reading any online reviews.
- Choose a square of blocks in your community and walk the perimeter. Don't do any research ahead of time. Notice the differences and similarities between landscape and buildings in the area. Take note of the kinds of people and animals you encounter.
- Commit to a week of assessing the weather without the help of digital devices.
- Commit to a number of days without your fitness tracker. Use a simple, self-made set of questions to assess your level of rest, tiredness, and physical exertion throughout the day.
- Leave your phone or tablet out of your bedroom for the night and try waking up to an alarm clock.
- Wear a watch to prevent looking at your phone for the time, avoiding seeing other message indicators when you do so.
- Turn off push notifications.
- Turn off message indicators. Determine a few preset times you will check all email, social media, etc. and do not check them any other time.

ASSESSING AND ADDRESSING "T" ISSUES: TIME

Time can be the most straightforward domain to measure when it comes to technology use. Complicating the ease of assessment, however, is the tendency of most of us to underestimate, sometimes by wildly large margins, the amount of time we interact with technology. To be frank, most of us spend much more time with our devices and in digital spaces than we like to admit. None of us likes to admit how many minutes we sink into wandering around the web or binging on Netflix. It doesn't feel good to come to terms with having spent two hours trying to accomplish a new level in a game we intended to engage for twenty minutes. It makes most of us feel uncomfortable when we look at the actual number of minutes we invest in our social networks, interest boards, games, and just plain old surfing.

For this reason, we need to get real and be honest. The best way to get to the truth of the numbers is by using one of the many apps on the market designed to accurately measure the time we spend on our smartphones, laptops, computers, and electronic tablets. These apps measure technology use and send reports regarding which sites/apps have been visited, how much time has been spent with each, and what times of each day. The data won't lie.

Whether we use digital tools or simply keep track on our own, tracking general amounts of time spent in each kind of digital spaces (work, entertainment, social media, etc.) can be very illuminative and also expose the patterns of our technology use. Are we often engaging technology when we are bored? Hungry? Tired? Do we use it as a filler when other activities might be more beneficial? Do we find ourselves without enough time for pursuits in which we'd like to engage simply because we've lost track of our time with devices? Do we use screens as we transition to sleep times and then have difficulty falling asleep? Do we engage them even before we've engaged our own minds when we wake up?

When we have an accurate estimate of the number of minutes spent with technology each day, we can multiply the number by seven and calculate the number of minutes spent with technology per week. Once we know the number, we need to look at that number in the context of the goals and hopes we have for our lives. We must ask ourselves if this amount of time feels acceptable to us and allows us to live a balanced, rich life that is moving toward our personal true north.

On its own, time is not a reliable measure. The reality is that time is really only a relevant domain when considered in relation either to the health of our embodied life or to the content with which we are interacting. If we are spending two hours a day with porn or highly violent content, that is likely having a less-than-neutral impact on us physiologically and emotionally.

If, however, we are spending two hours per day doing research online that will enable us to cure a disease, build a stronger building, or improve our community, we will find two hours not nearly enough. In other words, it's important to compare the time domain with the other assessment domains. For example, if the time spent on devices is significant but the content is intentionally educational or creative (e.g., schoolwork related to or for learning how to play the guitar or for creating electronic music), the high-quality content may balance out the time concern.

Being balanced is a key to assessing harmful versus healthy use. That is, by assessing the amount and nature of the time we spend with our devices and in digital spaces and then considering those findings alongside the other four domains, we can make informed and intentional choices about how to best structure our time to make space for healthy living. For example, if a new tech product is released that promises to make our lives easier, do we click on the "buy now" button, adding it to our already digitally driven lives? Or do we intentionally assess what the impact, positive or negative, of the new product will be and mindfully choose according to what is best for us personally?

We could all learn a lot from the way the Amish community evaluates whether to engage new technologies or not. Rather than simply buying into the cultural belief that new technology is always better than what existed before it, this careful community asks questions and allows for deep critical thought, heavily weighing the possible impact of a particular technology on their community as well as their youth. Professor Greg Crosby summarizes those questions:

1. **Would the technology need to be owned privately or might it be shared/borrowed?** When an entire community shares an object, there is a built-in need for waiting and communication. Not everyone can use something at one time. Sometimes a person or family will need to wait while someone is using the object. There will need to be communication between families and individuals to coordinate the schedule of use, and there will be a shared sense of responsibility for maintenance. All these factors result in opportunities for learning to delay, communicate, and act in accordance with what is best for a group. Imagine how video gaming might change if a neighborhood shared a single gaming system or two!

2. **How will engaging this technology change the family and/or community? Is it making us feel safe and secure or fragmented and distant?** For instance, if a house can be raised in two weeks without using any motorized or technology-based tools but can be raised in less than one week with those tools, what impact might this have on the community's sense of attachment and interdependence/reliance? Is any saved

time worth the potential cost? What does the community value more, interdependence or time?

3. By embracing a particular technology, what will the community also need to embrace to keep that technology up to date and effective? The consideration of latent effects is best undertaken *before* **the technology is engaged.** This question engages forethought about what future investments will need to be maintained once something new is added and provides important opportunities to consider the broader changes ushered in by even a single new device.

It strikes me that we might all be much better off if we spend even a small amount of time considering these questions. Rather than going from zero to one thousand by giving children who have not had their own phone a fully limitless smartphone, what about considering less immersive technologies to start with? Perhaps an old school Game Boy or a fully locked phone that only makes calls? Rather than find it necessary to keep updates current on a laptop, tablet, phone, and watch, why not consider which is most helpful and go with that alone for a while? Any time we ask ourselves questions that help us become more moderate, I believe we've gone a long way toward maintaining a healthy balance in our digital and embodied lives.

11

Spotting Points and the Fiery Life

There is so much beauty, wonder, and complexity to be experienced in the world, and helping people connect to those offerings is what drives my work. In truth, as both a person and a therapist, I care about technology only in relation to how it benefits and enriches relationships (with self and others) and how it might empower the living of complex, interesting, growth-mind-set-driven, rich, and fiery lives. But this is where my work is most challenging. It's one thing to do an assessment. It's another thing altogether to really consider the findings deeply and get active with the data. There are reasons we move through life quickly with little forethought; it is painstaking work to live with intention. It's even harder still when we need to break habits in order to do so.

As we discussed in chapter 9, "Habits and Norms," establishing healthy norms is easier than breaking bad habits. You might want to revisit that chapter and its contents now. The truth is that breaking habits is difficult work fraught with both victories and defeats. And this is particularly true with the habits we've evolved with our devices, especially because so many of them have garnered us status and a feeling of accomplishment or are truly required for us to participate in our work and social lives. Being the hyperaware, ever-informed, well-networked people that technology allows us to be has incredible benefits. These benefits make it seem counterproductive to moderate our device use. Not only is habit breaking hard, but breaking habits that we're not sure we want to break is hard times ten! To do so, we will need a clear vision of why the change is important and how we might make it happen.

Some "prep work" is needed before we begin to actively transform our engagement with technology in such a way that it becomes an accompaniment to and resource for living a well-rounded, fully engaged, embodied life. This preparatory period can make the process of breaking habits and resetting norms more manageable, while also providing space for truly committing to the process. Choosing the right time for the effort, rallying support, finding effective spotting points, and practicing mindfulness are all parts of the process.

THE RIGHT TIME FOR CHANGE

If we hope to be successful in changing a pattern or breaking a habit in our daily lives, we must choose the timing wisely. I often encounter people going cold turkey off technology after they have experienced cyberbullying or sexting, had difficult encounters with a clan in a MMORPG, or had real sexual side effects as a result of consuming massive amounts of porn. Understandably, nothing jolts us into action like a traumatic experience. In reality, however, these would be some of the worst times to try to make serious and lasting changes to our technology use habits.

For example, I would not recommend trying to reorient our relationship with technology when we are working through trauma, becoming new parents, moving, transitioning out of a major relationship, starting a new relationship, launching a new educational or vocational pursuit, trying a new fitness or lifestyle regimen, or engaging in any other kind of major life event. Any time we are already doing lots of new and focused work is not the time to try to change an entrenched habit, especially when that habit involves access to our social supports and important tools for our everyday living. We are going to need focused energy, space for practicing new skills, and the wiggle room for failure to transform the way we interact with technology. It will be hard work.

Choosing a time to make these changes and then sticking with it will require us to remember why we believe these changes are important. It can be helpful to take some time to write down the reasons that new norms are in order and refer to the list as the habit breaking commences. It can also be helpful to think through practical challenges that exist within the chosen time frame. Does a major holiday or personal event fall within the time frame? If it does, what kind of accommodations might help keep us realistic in our goal setting and habit breaking during this time? The more we can go into this time with our eyes wide open to the realities of our calendars, the better.

RALLYING SUPPORT

Informing others of the changes we intend to make is important. Over time, we have unconsciously trained the people around us to know and believe certain things about us. Today, this includes beliefs and expectations about how we will gather information and communicate. If we are avid social network users, we have likely trained our friends that we will be up on the happenings in their lives by following their profiles. If we are glued to our phones, responding quickly to every text, we have set an expectation with others that we are always reachable. In addition, if our professional work or the bulk of our social life happens within the digital domain, it is especially important to make certain to educate others about the changes we are making and to put systems in place for communication and meeting relational needs.

As we begin to shift our relationship with technology, we also will be shifting the way in which we connect with the people in our lives. This can be complicated and expose us to real consequences. We may feel adrift, lonely, irritable, and out of the loop as we work to moderate our tech engagement and begin to become less constantly informed regarding the happenings of others. We may miss out on information and sometimes on data that might really matter. This is especially true if we are involved in pursuits in which near-constant access to relevant data sent via digital means is crucial or if the bulk of our social time is spent with people in purely online spaces (e.g., MMORPGs or social networks). The sense of loss we might experience in this regard is real.

Given the shift in relationships that can occur when we choose less access to our devices, it is imperative that we do the difficult work of communicating clearly with those people who have expectations of us based on our past use. We must tell people we are backing off and prepare them that we will be less informed. We may need to ask them to inform us outside of social networks if real needs or issues arise. Just because we want to engage technology less doesn't mean our friends want to develop new ways of keeping us informed. We must talk with these people and determine ways we can remain connected apart from devices. We must listen to what they would like or need from us, and we must be willing to communicate our own wishes and needs as well. We also must keep these realistic. Just because we are breaking habits doesn't mean that everyone else wants to do the same. Graciousness from all parties and plenty of space for failures will benefit everyone involved.

Talking honestly *with ourselves* about the kinds of support we need to do the hard work of making changes can also be challenging. Do we need more physical activity to combat the agitation that will arise from resisting the impulse to check our devices? Do we need more embodied person-to-person contact? Do we need quiet? Books? Healthy food? A Rubik's Cube or some other handheld activities out and about the house? Do we need a hobby? To learn to knit? Take a class? Join the community center? Do we need images of beauty or objects related to a particular interest around us so that we aren't tempted to return immediately to the never-ending flow of images we can find on screens? Sometimes, finding support in like-minded countercultural people can help. So can inviting (not forcing) others to take the challenge alongside you!

FINDING OUR SPOTTING POINTS

Earlier in the book we discussed "spotting," the technique dancers use to maintain their balance and stay in place as they learn choreography. This skill requires dancers to find a fixed point off in the distance to which they can fix their gaze. Holding this spot in their visual field while they spin, snapping their head around so they maintain sight on the spot, allows them to avoid dizziness. When employed by a group of dancers, spotting keeps each dancer in a preset place during shared choreography. Artists and athletes of all kinds use similar tools. Snowboarders and skiers use colored lines in the snow to spot to as they jump, singers performing in large settings use in-ear monitors to help them stay on key, and character actors spot to the temperament and behavioral traits of the characters they play.

Each of us has certain things, people, attitudes, and ideas to which we spot. Some of these are set by us for ourselves; others are set for us by our families, culture, lifestyles, and vocations. For some of us, spotting points relating to roles we have filled, names we have been called, or situations we feel define us pull so strongly for our attention that it feels nearly impossible to orient our attention and energies to anything else.

In many cases, spotting points are the manifestation of external loci of control. In one of my favorite talks, I illustrate this point by describing a long list of names I have given myself or that have been given to me by others. These include but are not limited to names and roles such as daughter, mom, bully, friend, responsible, prude, stupid, loving, teacher's pet, liar, therapist, fat, strong, cute, and, my personal favorite, given to me at a talk in Northern Ireland, eejit.

During my talk, I write each of these words on a nametag and stick them on myself as I briefly describe the power behind each label or role. Once I am covered with "Hello my name is . . ." tags, I talk about how situations in my current life can inspire me to pull specific tags off, place them at a spot in front of me, and begin to spot to them. In so doing, I confirm over and over again that I am in fact stupid or . . . (any other of the tags). On a good day, I might spot to the label "loving." On a bad day, that would never, ever happen. On those days, I'm focused on spotting to words like "stupid," "fat," and the like. When I do this, as we all do, I create more evidence to support what I am already spotting to. The external locus of control is in full swing, and I am merely spinning in its orbit.

Technology is perhaps the most easy and accessible delivery system of spotting points today. Bereft of our own direction, we can easily find something of personal or completely impersonal relevance to spot toward. Given technology's constant dynamic nature and its ability to distract us, we easily allow the news, our friends' statuses, information, or even entertainment to create spotting points for our focus. We find information that makes us happy or sad and spot to it. We find bodies we believe ours should look like and spot to them. We notice parties we weren't invited to or honors we didn't achieve and focus relentlessly on those. We spot toward what we believe would make us valuable or smart or hip or whatever it is that captures our attention—until something else captures it. Even being well-informed (aka glued to our phones) can be a spotting point of sorts.

Rather than indiscriminately spotting toward whatever comes our way, we can identify and choose, *with intention*, the ideas, qualities, endeavors, pursuits, and traits that keep us on a grounded, healthy track in life. When chosen in concert with an honest assessment of who we genuinely are and supported by a growth mind-set model encouraging healthy risk-taking, these internally generated spotting points ground our living and open us to our fullest potential. The most fitting outcome—examining how our lifestyles, behaviors, habits, and norms either support or discourage the use of healthy spotting points—would be that of finding an internal locus of control that is strong enough to support a rich and fiery life. The question then becomes: "What are those internal spotting points that would compel us toward a reimagined and intentionally chosen, embodied life?"

SPOTTING TO THE FIERY LIFE

We have now arrived at the crux of this book: a challenge to discover the glorious benefits and mindful "how-tos" of living a life that feels and is worth the hard work of living it. Yes, life is complicated. If lived fully, it is rife with difficulties and pain as well as joy and wonder. Challenges pepper it, but opportunities do as well. Work, effort, and meaningful pursuit enhance the sweet parts of life, and a balance of exertion and rest seem to create the conditions for a sense of both competence and contentedness. When life presents struggles that are too heavy or prolonged in such a way that this balance is impacted, all manner of difficulties arise. The same is true if the balance is off in the direction of all play and no meaningful investment of energy. Striving to maintain our balance by spotting to our own true self is critically important. We must feel competent to complete meaningful tasks, to take appropriate risks, and to know when we need rest. We must know how to stimulate our bodies, minds, and hearts and how to soothe them as well.

We must also know what drives us, motivates us, and lights us up. Each of us has a unique way of being in the world, and few of us are spurred to action by the same stimuli. Words may provide the perfect motivation for some; statistics and their analysis may be required for others. Nature may move you to action; music may both calm and inspire you. Some people need a team setting in order to work effectively; others need solitude and quiet. There are many ways of being in the world. Once each of us has a sense of our own unique way of *being,* we can begin to identify the kinds of stimuli in the embodied world that will help us as we moderate our technology use.

I think of these person-specific motivators as "muses" that help people turn away from their technology. Muses, in Greek mythology, were personified forces who served as sources of inspiration for the gods. In my work, I often think of embodied experiences, tailored to a person's specific preferences and proclivities, as muses. If you are a creative soul, the opportunity to sculpt, paint, collage, or draw might be a muse to inspire less time in digital domains. If you are a person who comes alive in nature, hiking, birding, or rock collecting may be a muse to help you turn away from screens. For people who experience life with the body, yoga, acrobatics, massage, walking, jumping rope, juggling, or any such activity that gets their physical self engaged will be a welcome muse. Not all of us, however, are introduced to our muses automatically.

Unfortunately, if we have grown up lacking a goodness-of-fit between ourselves and our families, parents, community, or culture, our lives might feel bland or wildly out of balance. Perhaps we are artists raised by mathematicians who either belittle or undervalue the creative process. Perhaps we are mathematicians raised by circus performers who are completely baffled by our need for intricate and predictable structure. In either case, we might be left feeling completely out of touch with who we are genuinely to be and how we are to become. This lack of balance between who we are intended to be, what we are meant to do, and what we have been exposed to as options can leave us unable to recognize our muses. We truly might not know what gives us joy and what would be worth working hard to achieve. This leads to our struggling with how to live a rich, interesting, and meaningful life.

To find our muses, we might take cues from our tech life. For example, if we love playing racing games, we might find a racetrack in our area where we can drive a real race car, or we might build a go-kart. On a smaller scale, we might like to read exciting stories or drive with the windows down to feel a physical effect of our movement. If we love watching *Dancing with the Stars,* we might sign up for ballroom dance lessons. If we gravitate toward strategy games, we might find ourselves completely immersed by a game of solitaire with cards or in complex table games with others who like to play. If cooking shows dominate our screen time, a recipe book, ample ingredients, and a few hours in the kitchen may be even more fulfilling than the shows. Using where we travel online to help us find muses for exploration offline can be a real help.

It is important to note that a meaningful, fiery life must include opportunities for both exertion and rest. A rich, healthy, fiery life has, at its core, a grounding in unique purpose and moves neither too fast nor too slow. When we live fiery, balanced lives, we feel as though we are "in the groove," have "found our sweet spots," and feel "fully alive"—at least some of the time. We recognize a need for engagement in pursuits that aren't always only for ourselves, and we know we must attend to some tasks that are purely maintenance. When attempting something new, we do not need instant mastery because fiery living involves a full commitment to a growth mind-set model. This allows us to find opportunities to be deeply rooted as well as to stretch and grow into spaces that might be new to us. When we live like this, we take appropriate risks, seeing failure as a teacher and developing grit and resilience as we learn. Living a fiery life is far from tame, far from safe. But it is very, very good.

Fire is a complex element, as beautiful as it is dangerous. It has the power to purify and to destroy. When contained, it can create all kinds of wonder. When out of control, the damage can be immense. We can approximate fire with battery-operated candles, with screen images of crackling logs in a fireplace, or by filling a fire pit with construction paper and cellophane flames. All these will be safer than the real thing, but none will give us the heat or unpredictability of actual flames.

Giving our lives over to the internet or to the devices that deliver it to us is a bit like huddling up to a battery-operated candle to roast a marshmallow. This is not to say that some of our online, plugged-in experiences aren't fiery and dangerous or even life affirming and growth inducing. They can be, and they are. Without countering and interspersing these with embodied fiery experiences, however, we live out of balance and out of touch with the natural world of people and things around us. Doing deep work, having embodied experiences, feeling confident in our ability to be physically present in relationships and life, and maintaining healthy bodies all require exposure, at some point, to the real fire of our embodied world. To live a fiery life, then, we must know how to evaluate which kinds of fire are purifying and which are destructive.

Blacksmiths do this kind of discerning work with fire every day. Steeped in knowledge about the particular metal they use and the heat required to make it malleable, the blacksmith knows exactly when to subject a piece of metal to the fire. If we reduce their wisdom to a ridiculously simplified explanation, we can say blacksmiths work with a color scale, moving from red to orange to yellow to white in relation to the fire. A fire that is primarily red is not hot enough to effectively alter the molecular structure of the metal. A fire that is mostly white is too hot and will liquefy the metal. In between these two colors are orange and yellow. When a blacksmith's fire achieves the orange-yellow state, she can expose the metal, trusting the fire to make it malleable.

We can be blacksmiths in our own lives. If we only subject ourselves to experiences lacking in novelty or that are easy for us, like putting metal into red fire, we will not change much. If we place ourselves in situations that are too challenging, like putting metal into white fire, we will be overwhelmed and dysregulated. Discerning those experiences that produce the effects of orange-yellow heat is key, and our orange-yellow heat is unique to each of us. To know ourselves, find our spotting points, and realize our potential, we must be able to slow down, tolerate self-examination, and consistently practice the art of getting grounded—without distraction, without disruption, without devices.

TEN (RICH) MINUTES A DAY

Researchers at Stanford, Massachusetts General, and UCLA have found that ten minutes a day of mindfulness meditation for six months doubles the gray matter in the regions of the brain related to emotional well-being and executive function. This is amazing news. When I share it with people, they are excited and hopeful. Our brains can heal themselves; and on the way to doing this, we can learn to be still and get grounded and can strengthen our internal locus of control.

The trouble is that mindfulness meditation doesn't just happen. It must be learned. It must be practiced. The times we need it most are likely the times we cannot learn it, so we must make time for it long before we ever need to employ it as a mind-building, fiery, life-offering technique.

There are many ways to learn and practice mindfulness, contemplation, and other forms of meditative practice. For our purposes here, let's refer to whatever practice allows us to be fully aware of the present moment, using our breath along with the powers of directed attention to help us be still and openhearted. The most basic practice could look something like this:

A. Assume a posture of **Alert** restfulness. For many people, this is seated in a chair with both feet firmly on the floor. For others, it might be sitting on the floor or on a prayer or meditation cushion. Lying on your back is also fine. The goal, however, is restful alertness, not sleep.
B. Breathe. Try focusing on how it feels when breath enters and exits your body. Breathe in through the nose and out through the mouth ("smelling the roses and blowing out the candles") allowing your diaphragm to expand on the inhale and fall on the exhale. You can add words to your breathing if it helps. Try "releasing stress" on the exhale and "taking in space" on the inhale. You also can imagine your body

as a closed system. Any time you take something new into an already-filled closed system, something must be removed to make space for the new. As you breathe in spaciousness, you must release tension. Use your imagination to try to fill more than 50 percent of the closed system of your body with spaciousness.

C. Create space in your mind for simply being. As you focus on your breath, remind yourself that this ten minutes is simply for you to be within. There is nothing that needs your attention for the next ten minutes.

D. Direct distractions and **Draw** attention back to being. When you are beset with distractions, as we all constantly are, simply notice them, name them, and then do what you can to draw your attention back to your breathing. Many people find it helpful to use the analogy of a boat dock to practice directing distractions and redirecting the attention. If you imagine yourself as a boat dock, enjoying the feeling of the sun and the gentle rocking of the water, you can place distracting thoughts (tasks that need to be done, people to contact, feelings to feel, etc.) on boats that are far out in the water. Communicate to the boat that there is not space on the dock for it to moor right now. Then let it float by. If it is important, it will come back. Watch the boat floating by and, as quickly as you can, redirect your attention to being the dock. Practicing feeling the sun on the dock, feeling the rocking the water brings, and hearing the sounds you might hear will help you draw your attention back to being the dock. Bringing your awareness back to your breath can be an effective way of coming back to being as well.

Notice that these steps are ordered using words beginning with A, B, C, and D.

This is so we can remember the progression with relative ease. Once we have done this type of exercise several times, we will be able to attain a centered and calm stance more confidently. It is from this type of stance that we can check in with ourselves, determining where we are exposing ourselves to situations and ideas that will grow and mature us—and where we might be settling for less. This calm place of being is where our most robust spotting points can be discovered and where a strong internal locus of control is grounded.

12

Getting to Decide
(and Being Cheered on for Doing So!)

An apple seed will always produce an apple tree. It has no other choice: the type of seed always determines what kind of tree will grow from it. Many contextual elements, however, will impact a tree's growth. The richness of the soil in which it is planted, the amount of water and light it receives, the temperatures it must withstand, the winds it must endure—these conditions and more determine how healthy that tree will be, the depth and variety of the color of its leaves, the tastiness of its fruit. For example, in areas where strong winds blow consistently across the tree, it might take on a bent-over look—unless some wise gardener intentionally puts props in place prior to wind exposure.

The same is true for us. We are the product of our DNA, but we also are shaped by the contexts in which we grow. Were we offered opportunities for secure attachment? Did we experience trauma at some point in our lives? Were our sparks of interest nurtured? Have we practiced being empathetic enough? Is our curiosity regularly cultivated? Are we continuing to hone our social and emotional skills as we age? These elements and more of our "context" determine how healthy we will be, the richness of our lives, the joy of our days. For example, like the apple tree, if we encounter "strong winds," we might end up "bent over" or unhealthy.

Technology is just one example of a consistent strong wind that can affect the trajectory of our growth. Whether it shapes us in beneficial ways or harmful ones is mainly up to the exposure we allow and to the "props" we intentionally put in the place for support. Such props comprise the many tiny and large, mundane and complex, difficult and easy choices we make regarding technology and about our lives every day.

Over my many years of engagement with people, I am always humbled and thrilled by the honest efforts of brave individuals to make wise choices when it comes to interacting with their devices. I see adolescents and young adults fleeing social networks and reverting to flip phones in efforts to reduce distractibility and the temptation to live life online. I listen as gamers talk about wanting to develop worlds where pro-social and strategic elements in games aren't buried under violence. Developers are creating apps that can help special populations, that provide opportunities for support, and that turn off access to the internet and send passwords to resume use at set times. College professors are teaching important digital literacy skills alongside giving extra credit for "technology fasts," with students gobbling up the opportunity, finding a deep sense of peace after initial feelings of anxiety and depression. Medical students and engineers and programmers of all kinds use their screen time to develop crucial life-saving technologies, then relax by undertaking all manner of embodied experiences.

Every person has choices when it comes to technology engagement and balancing that with a full, fiery embodied life. We get to choose deep connection with the embodied alongside proficiency, skill, and discernment with the technological. We get to decide where our sense of self is grounded and how that same self connects to others. We get to demonstrate flexibility in our lives and exert our agency, embracing technologies that enhance our lives and moderating our use of those that don't. We can boycott platforms that hurt us or contribute to objectifying or injuring ourselves or others. We can push ourselves to keep our in-person communication skills intact and our ability to live through awkward moments well practiced. We can choose to know the essence of ourselves and our feelings, and we can regulate those feelings, even when they are inconvenient and uncomfortable.

By reading this book and getting to the end, you have demonstrated wholeheartedly your desire to consider and be intentional about your choices. For that, I celebrate you! I wish I were there with you in person so that I could make you a spicy meal or a fragrant cup of tea, then jump rope with you. Or perhaps we'd choose to climb a tree and watch a sunset or fly kites and then sit by a fire and sing some songs; I know someone who plays harmonica and someone else who plays guitar. Maybe you'll be the person who finally manages to teach me how to complete a Sudoku or who shows me how to make pasta. If you would please teach me how to Flamenco dance or to sing in Kiswahili, I would be beside myself with joy. And what may I teach you?

There are so many things I'd like to experience, and I'm guessing the same is true for you. Some of these experiences can be had online via devices and some cannot, at least not with the same effect. I believe, to my core, that a balance of these experiences is best and that finding that balance can be excruciatingly difficult. I also believe it is well worth the effort to try to find it at all costs. Your embodied life depends on it!

As we part, allow me to say that I hope my words have empowered you to break some habits, set some new norms, and gather fresh tinder for your fiery life. Be patient and consistent. And focus your efforts. Building a fire that can deliver orange-yellow flames that will burn as long as they need takes hard work and time. Above all, know that you can do it because you can *choose* to do it. I know you can.

Appendix

DISCUSSION QUESTIONS AND SUGGESTIONS
FOR FURTHER EXPLORATION

Regarding Embodiment

1. In what ways is my home welcoming to the entire body? How might I make it more sensually inviting?
2. What kinesthetic offerings do I have available (and out)?
3. What are two physical spaces that I might convert to screen-free spaces?

Regarding Physiology

1. What kinds of slower media/digital content might I use to replace faster-moving content?
2. How can I enhance my own self-soothing skills and teach them to those around me? Get very specific about this.
3. How might I force delay of gratification some of the time? What actions can I take to practice doing one thing at a time?
4. How might I create and celebrate opportunities for boredom?
5. What kinds of violent/sexualized/monetized media do I use without thinking about it? Is there a replacement that includes the elements I like (e.g., strategy, complexity, quality of design) that omits the negative elements?

Regarding Relationships

1. Do I interact with others verbally? Do I make eye contact?
2. In what ways do I model judgment of others or "shopping" and objectifying others?
3. How do I feel when I engage social media? Do I compare myself? Do I experience FOMO? If so, how do I respond to it?

4. Do I regularly engage digital content that actively objectifies others?

5. What kind of relational aggression have I witnessed in digital spaces? How did it make me feel?

Regarding the Development and Health of the Sense of Self

1. Do I model/teach/value internal locus of control?

2. Do I allow (invite) failures in and for myself and for others? Can I tolerate the discomfort?

3. Do I tolerate and make space for boredom? If not, how might I do this practically?

4. How can I enhance my ability to focus? To delay? To regulate?

5. What notifications could I turn off right now?

Technology Pyramid Worksheet

List the devices, apps, software, games, and websites you interact with on a daily basis, grouping them under:

Technologies that...

Connect | Educate | Entertain | Hurt

Imagine the triangle below represents 100% of your technology time in a day. Proportion out the triangle below to represent how much time you spend with each of the technology categories. The technology category you spend least time with will go at the top and that with which you spend most time will go at the bottom.

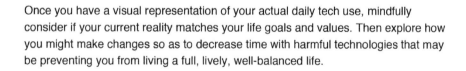

Once you have a visual representation of your actual daily tech use, mindfully consider if your current reality matches your life goals and values. Then explore how you might make changes so as to decrease time with harmful technologies that may be preventing you from living a full, lively, well-balanced life.

For a downloadable worksheet you can fill out, go to https://rowman .com/ISBN/9781538115848/Deviced!-Balancing-Life-and-Technology -in-a-Digital-World

Technology Assessment Chart

Ask yourself, "How am I doing today/this month, when it comes to my technology engagement?"

| Name / Date: _____

	Ability to Focus	
A	**Ability to Delay**	
	Ability to Regulate	
B	**│ Balance**	
C	**│ Content / Context**	What broad category of technology do you gravitate toward? Strategy, Logic, Social/Relational, Media/Pop Culture, Creative Arts, Body/Kinesthetic
D	**│ Devotion**	
T	**│ Time**	

Take time to consider where you are in relation to these categories that might be issues when it comes to technology use. You might use the numbered scale below to describe your tech engagement in each domain (or that of whomever you are doing the assessment on behalf of). If short narrative notes work better for you, do that instead.

1 - Things are going well for me here. I feel I have accomplished this.
2 - I'm doing OK here but need a little work.
3 - This is an area I need to work on.
4 - I am not doing well here. I need to put in major work to master this.
5 - I have so much work to do here that I'm not even close to mastery.

Returning to this assessment on a regular basis can be helpful as you mindfully attempt to break some habits, set some new norms, and gather fresh tinder for your fiery life.

For a downloadable chart you can fill out, go to https://rowman.com/ISBN/9781538115848/Deviced!-Balancing-Life-and-Technology-in-a-Digital-World

EMBODIED ACTIVITIES AND RESOURCES TO TRY IF YOU GRAVITATE TOWARD STRATEGY

Books

Chasing Vermeer
Calder Game
Fantasy series
Great Art Detective series
Mystery series
Science fiction series

Documentaries

Enron: The Smartest Guys in the Room
Hard Problems
Pressure Cooker
The King of Kong
We Steal Secrets

(Non-Digital) Games

7 Wonders
Battleship
Bears vs. Babies
Blokus
Checkers
Chess
Dominion
Exploding Kittens
Forbidden Desert
Forbidden Island
Hoot Owl Hoot
Magic
Mahjong
Mysterium
Pandemic
Rat a Tat Cat
The Resistance
Risk
Rummikub
Saboteur

Sequence
Settlers of Catan
Spot It!
Stratego
Tic-tac-toe
Ticket to Ride
Worst Case Scenario

Podcasts

Decode DC
Planet Money
Pod Save America
So That Happened
Stuff You Should Know

Assorted Embodied Ideas

Bonsai tree pruning
Bouldering
Budgeting activities/stock market exploration
Camp games (Capture the Flag)
Cooking/baking
Escape rooms
Geocaching
Jujitsu
Model rocket construction
Neighborhood or city scavenger hunts with clues
Number tiles games (putting the numbers in order)
Orienteering
Origami
Paper-airplane folding and flying
Pentominoes
Rubik's Cube
Robotics
Sudoku
Team sports (including robotic teams)

EMBODIED ACTIVITIES AND RESOURCES TO TRY IF YOU
GRAVITATE TOWARD LOGIC

Books

   Clue series
   FBI thrillers/mysteries
   Historical fiction and novels
   Miss Fisher's murder mysteries
   Mysteries (by Agatha Christie, Donna Andrews, Janet Evanovich, Sue
      Grafton, J. A. Jance)
   Poetry
   Stories that focus on ethics/philosophy
   Theory books

Documentaries

   The Examined Life
   The Parking Lot Movie
   Philosophy Kings
   Undefeated
   WordPlay

(Non-Digital) Games

   Bazaar
   Boggle
   Countdown Cars
   Clue
   Forbidden Desert
   Forbidden Island
   Pallina
   Rat a Tat Cat
   Riddles with Friends
   Rush Hour
   Scrabble
   Set
   Solitaire
   Sudoku
   Suspend
   Super Fight
   Traffic Jam

Word Brain
World Problems
Would You Rather

Podcasts

Deadly Manners
Intelligence Squared
My Favorite Murder
Philosophy Talk
Planet Money
Radio Lab
Serial
The Public Radio Alliance

Assorted Embodied Ideas

Ballroom dance
Building toys (Legos, K'Nex, blocks)
Cause-and-effect experiments
Coding
Etch A Sketch
Freestyle rap
Gratitude practice
Labyrinths (finger and walking)
Lateral thinking puzzles
Mazes
Meditation
Music (many genres)
Pentominoes
Personality assessments
Puzzles
Repair training (auto, bikes, home)
Science experiments
Shakespeare plays
Shape/pattern blocks

EMBODIED ACTIVITIES AND RESOURCES TO TRY IF YOU GRAVITATE TOWARD SOCIAL/RELATIONAL CONNECTION

Books

Autobiographies/biographies
Character-driven stories
History/historical event-themed books
Memoirs
Personality theory/psychology/sociology/anthropology
Self-help/relational books

Documentaries

20 Feet from Stardom
30 Days (Morgan Spurlock series)
A Touch of Greatness
Afghan Star
The Beauty Academy of Kabul
Bill Cunningham: New York
Brooklyn Castle
Catfish
Craig's List Joe
The Eyes of Tammy Faye
Extremely Loud and Incredibly Close
Finding Kind
First Position
The Hobart Shakespearians
I Am Big Bird
Jiro Dreams of Sushi
Paul Williams Still Alive
POV documentaries
Queen of Versailles
Skateistan
Stories We Tell
Straightlaced: How Gender's Got Us All Tied Up
The Way We Get By
Undefeated

(Non-Digital) Games

   Anomia
   Apples to Apples
   Catch Phrase
   Celebrity
   Code Names
   Committee
   Dixit
   Fluxx
   Go Fish Yourself
   Guestures
   Hear Me Out
   Imagine If . . .
   Most Likely to . . .
   Pandemic
   Red Flags
   Scum
   Secret Hitler
   Soul Pancake
   Stand Out
   Telestrations
   Wink
   Would You Rather

Podcasts

   The Cult of Pedagogy
   Invisibilia
   The Life Coach School
   The Liturgists
   The Memory Palace
   Mindful Mommas
   The Monti
   The Moth
   On Being
   Thrilling Adventure Hour
   Personality Hacker
   Professor Blastoff
   Revisionist History
   Scene on Radio
   Selected Shorts
   Story Corp

TED
The Truth
Theory of Everything
This American Life
Viceland Channel
What's Good with Stretch and Bobbito
Where Should We Begin

Assorted Embodied Ideas

Anthropological exploration
Camp games (sardines, obstacle courses, rope courses, human knot, shoe shuffle, etc.)
Organize yourself according to . . . ten things in common, bigger/better, etc.
Escape rooms
Civic events/festivals
Drum circles
Facetime/Skype
Improv classes
Interplay
Live music and theater events
Meetups
Partner yoga and acrobatics
People scavenger hunts
Personality assessments (Meyers-Briggs, Enneagram, Strength Finder)
Photo booths
Shared activities in a group (painting, classes, game/cooking nights)
Sidewalk chalking (plus streets and driveways)
Team activities
Theater classes
Volunteer opportunities in a group

EMBODIED ACTIVITIES AND RESOURCES TO TRY IF YOU GRAVITATE TOWARD MEDIA AND POP CULTURE

Books

History-themed books
Period-piece books
Sci-fi and technology-driven plots such as *Ready Player One*

Documentaries

5 Broken Cameras
Afghan Star
America the Beautiful
The Corporation
Enron: The Smartest Guys in the Room
Human Flow
Ken Burns's documentaries
King Corn
Lo and Behold
Miss Representation
Nursery University
Page One: The New York Times
Pom Wonderful: The Greatest Movie Ever Sold
POV documentaries
Searching for Sugarman
Screenagers
September Issue
The World According to Sesame Street
This Is Not a Movie

(Non-Digital) Games

Jeopardy! (on Alexa or with people)
Mystery Date
Random Facts
Trivial Pursuit

Podcasts

2 Dope Queens
All Things Considered
Decode DC
Freakonomics
How Did This Get Made?
On the Media
Minimalist
Planet Money
Pop Culture Happy Hour
So Many White Guys
What Do You Mean?
Whose Line Is It Anyway?
Viceland Channel

Assorted Embodied Ideas

Coding
Field trips to television/radio/tech locations
Filmmaking/editing
Letters to the editor/op-ed pieces
Live theater/dance/music/improv/sports
Magazine/newspaper subscriptions
Media and pop culture festivals
Museums
Political exploration
Stop-motion filmmaking
Website construction

EMBODIED ACTIVITIES AND RESOURCES TO TRY IF YOU GRAVITATE TOWARD THE CREATIVE ARTS

Books

Art and Fear
The Artist's Way
Biographies of artists
Comics
Graphic novels
Picture/coffee table books
Poetry

Documentaries

A Man Named Pearl
Ai Weiwei
Exit through the Gift Shop
First Position
Jiro Dreams of Sushi
Kings of Pastry
Louder than a Bomb
Manufactured Landscapes
Muscle Shoals
Pina
Samsara
Searching for Sugarman
Wasteland

(Non-Digital) Games

    Charades
    Dixit
    Dictionary
    Word association

Podcasts

    99% Invisible
    A Piece of Work with Abbi Jacobson
    The Monti
    The Moth
    This American Life
    Selected Shorts
    Story Corp

Assorted Embodied Ideas

    Adobe Illustrator classes
    Art for Kids Hub
    Art museums/galleries (some also have amazing websites)
    Basic drawing/sketching class
    Body paints/henna
    Buddha boards, chalkboards, or mural walls in the home
    Cake decorating
    Camp games (Pictionary on people's backs)
    Card making
    Collaging
    Cooking/baking
    Conserva and EdX art classes
    Crafting (with emphasis on creating, not on end product)
    Clay, playdough, blocks or other things to build with
    Dance or creative movement
    Gardening/landscaping
    Felting or other fiber arts
    Improv classes
    Interplay
    Lettering/calligraphy
    Live theater/dance/music
    Music lessons
    Needle crafts
    Origami
    Quilting

Paper crafts
Photography
Poetry readings
Sewing
Sidewalk chalking
Slam poetry
Soul collage

EMBODIED ACTIVITIES AND RESOURCES TO TRY IF YOU GRAVITATE TOWARD BODY-RELATED INTERESTS

Books

Anatomy and physiology books
Biographies/autobiographies of artists and athletes
Instructional books (sports, yoga, skill-based hobbies)

Documentaries

Pina
Rivers and Tides
Strictly Ballroom
Touch the Sound
Undefeated

(Non-Digital) Games

Bop It
Field Games
Funny Bones
Go Fish Yourself
Jacks
Jenga
Musical Chairs
Ninja Assassin
Quelf
Sardines
Shuffle Buns
Spoons
Suspend
Twister

Podcasts

   Hidden Brain

Assorted Embodied Ideas

   Aerial arts
   Aerial yoga
   Balance board
   Bouldering/rock climbing
   Bubble baths
   Circus arts
   Clay/playdough
   Cooking/baking
   CrossFit
   Cup stacking
   Dance lessons
   Driving (rent a racetrack)
   Drum circles/lessons
   Etch A Sketch
   Essential oils
   Fidget spinners
   Fitness classes
   Gardening
   Geocaching
   Gyro rings
   Hiking
   Hoberman spheres (breathing balls)
   Hula hoops
   Ice blocking
   Incense
   Improv classes
   Jumping rope
   Juggling (balls or sticks)
   Kendama (or other skill toys)
   Kinetic Sand
   Knitting/crocheting
   Legos
   Light specific to different settings
   Live-action role-playing
   Martial arts
   Meditation
   Monkey bars

Nature walks
Nerf guns/Nerf balls
Origami
Perplexus balls/mazes
Rituals/rites of passage
Scootering
Sensory stimulation (smells, tastes, sounds)
Sidewalk chalking
Sign language lessons
Skateboarding
Skating (ice, roller, or derby)
Snow sports (sledding, boarding, skiing, shoeing)
Spikeball
Stretching
Swing
Table tennis
Tap dancing
Tumbling/acrobatics
Weightlifting
Yoga
Yoga balls (and other alternative seating)
Yo-yos

Selected Bibliography

American Academy of Pediatrics (AAP). "Cell Phone Radiation and Children's Health: What Parents Need to Know." HealthyChildren.org (web page of AAP). Last modified June 13, 2016. Accessed January 24, 2018. https://www.healthy-children.org/English/safety-prevention/all-around/Pages/Cell-Phone-Radia-tion-Childrens-Health.aspx.

Ansari, Aziz, and Eric Klinenberg. *Modern Romance*. New York: Penguin Books, 2016.

Bedrosian, T. A., and R. J. Nelson. "Timing of Light Exposure Affects Mood and Brain Circuits." *Translational Psychiatry* 7 (January 31, 2017): e1017. doi:10.1038/tp.2016.262.

Borsellino, Lisa. "Exposure to Violence in the Media Leads to Aggression." *Medical News Bulletin*, August 30, 2017. https://www.medicalnewsbulletin.com/exposure-violence-media-leads-aggression/.

boyd, danah. *It's Complicated: The Social Lives of Networked Teens*. New Haven, CT: Yale University Press, 2014.

Carr, Nicholas. *The Glass Cage: How Our Computers are Changing Us*. New York: W. W. Norton and Company, 2015.

———. *The Shallows: What the Internet Is Doing to Our Brains*. New York: W. W. Norton and Company, 2011.

Chödrön, Pema. *The Wisdom of No Escape and the Path of Loving Kindness*. Boulder, CO: Shambhala Publications, 2001.

Czeisler, Charles A. "Perspective: Casting Light on Sleep Deficiency." *Nature* 497, no. 7450 (May 23, 2013). doi:10.1038/497S13a.

Dunckley, Victoria L. *Reset Your Child's Brain: A Four-Week Plan to End Meltdowns, Raise Grades, and Boost Social Skills by Reversing the Effects of Electronic Screen-Time*. Novata, CA: New World Library, 2015.

Dweck, Carol S. *Mindset: The New Psychology of Success: How We Can Learn to Fulfill Our Potential*. New York: Ballentine Books, 2007.

Gianluca, Tosini, Ian Ferguson, and Kazuo Tsubota. "Effects of Blue Light on the Circadian System and Eye Physiology." *Molecular Vision* 22 (2016): 61–72.

Giedd, Jay N. "The Digital Revolution and Adolescent Brain Evolution." *The Journal of Adolescent Health* 51, no. 2 (2012): 101–5. doi:10.1016/j.adohealth.2012.06.002.

Hamilton, Jon. "Heavy Screen Time Rewires Young Brains, for Better and Worse," *Shots: Health News from NPR* (November 19, 2016). Accessed January 7, 2018.

https://www.npr.org/sections/health-shots/2016/11/19/502610055/heavy-screen-time-rewires-young-brains-for-better-and-worse.

Hölzel, Britta K., et al. "Mindfulness Practice Leads to Increases in Regional Brain Gray Matter Density." *Psychiatry Research: Neuroimaging* 191, no. 1 (January 30, 2011): 36–43. doi:10.1016/j.pscychresns.2010.08.006.

Kardaras, Nicholas. *Glow Kids: How Screen Addiction Is Hijacking Our Kids—and How to Break the Trance.* New York: St. Martin's Press, 2016.

Kosenko, Kami, Geoffrey Luurs, and Andrew R. Binder. "Sexting and Sexual Behavior, 2011–2015: A Critical Review and Meta-Analysis of a Growing Literature." *Journal of Computer Mediated Communication* 22, no. 3 (May 15, 2017): 160. doi:10.1111/jcc4.12187.

Loh, Kep Kee, and Ryota Kanai, "High Media Multi-Tasking Is Associated with Smaller Gray-Matter Density in the Anterior Cingulate Cortex." *PLOS One* (September 24, 2014). doi:10.1371/journal.pone.0106698.

Malady, Matthew J. X. "The Useless Agony of Going Offline." *The New Yorker,* January 27, 2016. Available online at https://www.theatlantic.com/magazine/archive/2017/09/has-the-smartphone-destroyed-a-generation/534198/.

Masarie, Kathy, et al. *Face to Face: Cultivating Kids' Social Lives in Today's Digital World.* Portland, OR: Family Empowerment Network, 2014.

Newport, Cal. *Deep Work: Rules for Focused Success in a Distracted World.* New York: Grand Central Publishing, 2016.

The Nielsen Company. *The Nielsen Total Audience Report: Q1 2016.* Accessed January 7, 2018. http://www.nielsen.com/content/dam/corporate/us/en/reports-downloads/2016-reports/total-audience-report-q1-2016.pdf.

Rideout, Vicky. *The Common Sense Census: Media Use by Tweens and Teens.* San Francisco: Common Sense Media, 2015. Accessed January 7, 2018. https://www.commonsensemedia.org/research/the-common-sense-census-media-use-by-tweens-and-teens.

———. *The Common Sense Census: Media Use by Kids Ages Zero to Eight.* San Francisco: Common Sense Media, 2017. Accessed January 7, 2018. https://www.commonsensemedia.org/research/the-common-sense-census-media-use-by-kids-age-zero-to-eight-2017.

Rosin, Hanna. "Why Kids Sext," *The Atlantic,* November 2014. Available online at https://www.theatlantic.com/magazine/archive/2014/11/why-kids-sext/380798/.

Siegel, Daniel. *Pocket Guide to Interpersonal Neurobiology: An Integrative Handbook of the Mind.* New York: W. W. Norton, 2012.

Stein, Joel. "Why We're Losing the Internet to the Culture of Hate: How Trolls Are Ruining the Internet," *Time,* August 18, 2016. Available online at http://www.time.com/4457110/internet-trolls/.

Turkle, Sherry. *Alone Together: Why We Expect More from Technology and Less from Each Other.* New York: Basic Books, 2011.

———. *Reclaiming Conversation: The Power of Talk in a Digital Age.* New York: Penguin Books, 2015.

Twenge, Jean M. "Have Smart Phones Destroyed a Generation?" *The Atlantic,* September 2017. Available online at https://www.theatlantic.com/magazine/archive/2017/09/has-the-smartphone-destroyed-a-generation/534198/.

———. *iGen: Why Today's Super-Connected Kids Are Growing Up Less Rebellious, More Tolerant, Less Happy—and Completely Unprepared for Adulthood—and What That Means for the Rest of Us*. New York: Simon and Schuster, 2017.

RECOMMENDED WEBSITES

commonsensemedia.org: Reviews media for children and parents. Also provides digital literacy curriculum and up-to-date information on current trends in new media.

emergent.info: Real-time rumor tracker based out of The Tow Center for Digital Journalism at Columbia University.

fightthenewdrug.org: Research and resources regarding porn.

thatsnotcool.com: Hip, well-constructed, and reliable resources for healthy tech use for adolescents and adult allies.

twloha.org: To Write Love on Her Arms is a resource for individuals dealing with depression, anxiety, addiction, self-injury, and suicide.

Index

About the Author

doreen dodgen-magee, PsyD, is a psychologist with a private practice in Portland, Oregon. She maintains an international speaking docket and is followed online via her blog and Instagram, where she posts challenges for living moderately with tech and wildly in embodied spaces. One of Doreen's primary passions is engaging people about how the new digital landscape is shaping humanity. Far from being technology phobic or averse, she inspires the consideration of potential new norms around device use and encourages people to become moderate, not abstinent, in their tech engagement.

On a personal note, Doreen is the mother by birth of two young adults and "extra mother" by choice of many others. One of her most important roles is that of Auntie. She is a gun violence prevention activist, an Everytown for Gun Safety Survivor Fellow, and a Quaker. She loves going barefoot, drinking black coffee, the color red, dancing, documentaries, her meditation cushion, and people, especially people. She and her husband live in Portland, Oregon, where they practice radical hospitality and enjoy partaking in all that is their wonderful city.

Lightning Source UK Ltd.
Milton Keynes UK
UKHW010843141218
333944UK00006B/133/P